De Havilland Aircraft of World War I
Volume 2: D.H.5 – D.H.15

Colin Owers

Color Illustrations by
Juanita Franzi

Flying Machines Press • Boulder, Colorado

De Havilland Aircraft of World War I, Vol. 2: D.H.5 - D.H.15
by Colin Owers

Copyright © 2001 by Colin Owers

ISBN 1-891268-18-X
Printed in the United States of America

Published by Flying Machines Press, a division of
Paladin Enterprises, Inc.
Gunbarrel Tech Center
7O77 Winchester Circle
Boulder, CO 80301, USA
+1.303.443.7250

Direct inquiries and/or orders to the above address.

All rights reserved. Except for use in a review, no portion of this book may be reproduced in any form, including the internet, without the express written permission of the publisher.

Publisher's Cataloging-in-Publication
(Provided by Quality Books, Inc.)

Owers, Colin, 1944–
    De Havilland aircraft of World War I / Colin Owers ;
color illustrations by Juanita Franzi. -- 1st ed.
    v. cm.
    Includes bibliographical references.
    CONTENTS: vol. 1. D.H.1-D.H. 4 -- vol. 2.
D.H.5-D.H.15.
    ISBN 1-891268-17-1 (vol. 1)
    ISBN 1-891268-18-X (vol. 2)

    1. Airplanes, Military--France--History.
2. De Havilland aircraft--History. 3. World War, 1914-1918--
Aerial operations. I. Title.

UG1245.F8O94 2001   623.7'461'094409041
                    QBI01-700620

Neither the author nor the publisher assumes any responsibility for the use or misuse of information contained in this book.

Book and cover design, layout, and typesetting by John W. Herris.
Color aircraft illustrations by Juanita Franzi.
Cover painting "Into the Lion's Den" by Michael O'Neal.
Aircraft scale drawings by Colin Owers.
Digital scanning and image editing by Aaron Weaver and John W. Herris.

Visit our Web site at www.flying-machines.com

Table of Contents

D.H.5 1

D.H.6 11

D.H.9 17

D.H.9A 34

USD-9A 45

D.H.10 Amiens 47

D.H.11 Oxford 50

D.H.14 Okapi 51

D.H.15 Gazelle 52

Drawings 53

Colors and Markings 74

Color Plates 77

Acknowledgments

This book would not have been written without the help of many people. Special thanks go to Jack Bruce, that doyen of World War I aero historians, whose researches have formed the basis of this book. Jack has also provided photographs from the J.M. Bruce/G.S. Leslie collection as well as the Air Department drawing of the D.H.1 which formed the basis of my plan.

In addition photographs have been made available by the following: RAAF Museum, RDAF Museum, R. Gentilli, F. Gerdessen, T. Goworek, P.H.T. Green, P.M. Grosz, N. Hayes, J. Hopton, M. Krkyzan, D. McGuiness, the George Williams collection of the University of Texas in Dallas, USAF Museum, and US National Archives. The rest are from the author's collection. Where known the original World War I veteran from whose collection the photograph was copied is credited.

Special thanks to my wife, Julie, who typed the manuscript. Any errors or omissions are mine alone.

De Havilland

Geoffrey de Havilland built his first aircraft with the assistance of his wife and his friend, F. T. Hearle, and £1,000 contributed by his grandfather, in 1908–1909. This and his second aircraft built in 1910, was powered by a 45 hp engine of his own design built by the Iris Motor Company. He then designed four aircraft for the government after his appointment as designer at H.M. Balloon Factory (later the Royal Aircraft Factory). From these beginnings he joined the Aircraft Manufacturing Co., Ltd., where his first designs were the D.H.1 and D.H.2, classic twin-boom pusher biplanes.

Volume 2 continues the story with de Havilland's attempts to produce a tractor fighter, the D.H.5; a safe trainer, the D.H.6; and his greatest failure, the D.H.9 bomber. The final production effort was the superlative D.H.9A which saw limited combat use in the war but had a long and successful post-war career with the RAF and Empire air forces. The D.H.10 day bomber was just too late for hostilities but soldiered on in the Middle East. A sprinkling of experimental types rounds out this volume.

D.H.5

In the D.H.5 de Havilland sought to maintain the pusher scout's excellent view with the synchronized machine gun layout of a tractor scout. To achieve this the D.H.5 was rigged with 27 inches (686 mm) of back stagger to bring the pilot in front of the upper mainplane. Unfortunately, this meant that he had no rear view. The prototype, A5172, was otherwise a conventional biplane with a wooden box girder fuselage, wire braced with ply strengthening on the forward end. The 110 hp Le Rhone rotary was sharply faired into the angular box like rear fuselage. The main fuel tank was placed directly behind the pilot with a gravity tank on top of the starboard mainplane. Mainplanes were the usual two spar type with the spars spindled out between the ribs. Armament comprised a single fixed Vickers gun on the forward fuselage within easy reach of the pilot.

Production D.H.5 aircraft were distinguished from the prototype by their unbalanced rudders and faired fuselages which tapered from circular at the engine mount to octagonal at the tail. The type entered service in France in the spring of 1917 with Nos. 24 and 32 Squadrons, the former receiving its first D.H.5 on 1 May 1917. Deliveries were slow with the two squadrons operating their D.H.2 pushers alongside the D.H.5.

No. 32 suffered a number of casualties during this transition period due to the peculiarities of the D.H.5. Eighteen year old 2/Lt. Roland V. Williams was killed on 6 June in a flying accident; 2/Lt. K.G. Cruikshank was injured in a crash on the 17th, with Lt. St.C. Tayler and 2/Lt. L.W. Barnet suffering accidents on 24 June and 3 July, respectively. Once it was mastered the D.H.5 proved exceptionally strong, easy to fly, and fully aerobatic. However, it was easily outperformed by contemporary fighter aircraft above 10,000 feet (3,050 m). Relegated to ground attack duties, it performed sterling service in the Battle of Ypres in August 1917 and the Battle of Cambrai.

Proof of its sturdiness was given on 16 November 1917, when 2/Lt. W.R. Jones of No. 32 Squadron had an anti-aircraft shell pass completely through the fuselage of A9269. Only minor damage was sustained and he landed safely. Fortunately, the squadron began to receive S.E.5a fighters early in 1918.

In addition Nos. 41, 64, and 68 (Australian) Squadrons were equipped with the type. According to the Official History of the Australian Flying Corps, the Air Ministry issued instructions for a low flying "trench-strafing" aircraft which was to give the pilot a good view overhead as well as downward and in front. The Australians were soon to put their D.H.5 fighters

The prototype D.H.5 had a slab-sided fuselage, the cowl being abruptly tapered into the fuselage. The back stagger of the wings is evident in this view. (Credit: J.M. Bruce/G.S. Leslie)

to good effect as "trench-strafers." Even with this mediocre equipment they managed to achieve a number of aerial victories.

No. 68 (Australian) Squadron arrived at Baizieux, France, on 21 September 1917, with 15 aircraft. The squadron's complement of 18 fighters was made up from 2 Aeroplane Supply Depot. These were arranged into three flights as follows:

Serial	Identification Letter/Numeral	Usual Pilot
"A" Flight		
A9284	A	Lt. Richard W. Howard
A9273	B	2/Lt. J.R.Y. Bartlam
A9245	C	Lt. Les H. Holden
A9459	D	Capt. Wilfred A. McClaughry
A9226	E	2/Lt. C.H. James
A9399	F	Lt. Leslie N. Ward
"B" Flight		
A9464	1	Capt. Gordon C. Wilson
A9265	2	Lt. Archie J. Pratt
A9288	3	Lt. Roy C. Phillips
A9462	4	Lt. Fredrick G. Huxley
A9224	5	Lt. Harry Taylor
A9263	6	2/Lt. C.C. Sands
"C" Flight		
A9469	U	Lt. Albert Griggs
A9473	V	Capt. John Bell
A9242	W	Lt. Douglas G. Morrison
A9457	X	Capt. George C. Matthews
A9483	Y	Lt. William A. Robertson
B377	Z	Lt. Robert W. McKenzie

The Squadron had its first aerial engagement on 2 October. This was the first aerial encounter between an Australian squadron and the enemy in France. Four aircraft of "A" Flight dived on a German two-seater, which, in spite of the Australian's superior height, was able to escape due to its superior speed. Fifteen minutes later another two-seater escaped the same way. The Flight returned to base without Lt. Ivor Agnew who was forced to land behind enemy lines in A9271 to become a POW. He had engaged in both attacks and it was presumed that his D.H.5 had been damaged in the short engagement. On the 13th Morrison was lost. Returning from patrol, Morrison was observed some distance to the rear of the Flight of five D.H.5 fighters and McKenzie, his "fighting partner," went back to pick him up when they were attacked by four Albatros scouts. McKenzie flew straight at the leader, firing a burst at long range but was forced to withdraw with engine trouble. Although he could not gauge the effect of his firing, McKenzie was allowed to escape although his aircraft was virtually crippled. Morrison was shot down to crash between the lines. He was rescued by men from the 13th London Regiment before his machine was shelled by the enemy. Unfortunately, he died shortly thereafter.

The squadron had eight crashes in October which required replacement machines, none of which were due to enemy action. Engine problems continued to plague the D.H.5.

The Battle of Cambrai saw Nos. 64 and 68 Squadrons assigned the role of ground attack. The D.H.5 was fitted with a bomb rack which could carry four 25 lb. (11 kg) bombs. The morning of 20 November saw all three Australian Flights take off. Because of the fog and rain low flying in flight formation was impossible and the aircraft split up into pairs, but were soon acting individually. Capt. John Bell in A9473 was probably the first loss. Shot through the chest he crashed, was rescued, but succumbed to his wounds. Lt. Robertson was shot up by an enemy aircraft and landed A9483 at the advanced landing ground at Bapaume.

Huxley reported that he had dropped one bomb plumb upon a gun carriage, machine gunned the men around it, and killed three horses. He then blocked the road into Cambri by smashing a supply wagon in a similar manner. Flying through the mist he saw a body of 300 men, drawn up in fours as if on parade. "This parade was dismissed quicker than parade ever was before."

A9449 was the personal machine of Capt. Gordon C. Wilson, MC, AFC, DCM. Wilson is shown seated in the aircraft near Baizieux in December 1917 just before it was handed over to No. 2 Air Depot when No. 68 (Australian) Squadron re-equipped with S.E.5a fighters. The numeral "1" denotes that the aircraft belongs to "B" Flight while the streamers attached to the interplane struts indicates Wilson's status as Flight leader. The thin white band to the rear fuselage is the squadron marking. Also worthy of note are the telescopic Aldis sight and bomb rack under the fuselage. Received on 28 October 1917, A9449 served with distinction until 19 December. Wilson earned his MC during the Battle of Cambrai on this aircraft. (Credit: G. Muir)

The presence of an Avro 504 (A593) and a Curtiss Jenny in the background indicates that this D.H.5 (A919?) is a training machine. This aircraft falls into batch A9163–A362 ordered from the Aircraft Manufacturing Co. under Contract 87/A/1286. The last aircraft from this contract for 200 machines was cancelled and the prototype substituted instead.

D.H.5 B365 was one of 50 machines (B331–B380) manufactured by the British Caudron Co. Ltd. under contract 87/A/1433. The access to the oil and fuel filler tanks in the headrest can be seen. The gravity tank's position is well displayed. This is thought to be another training machine.

A9242 was the second D.H.5 to bear the inscription *New South Wales No. 14 Women's Battleplane.* It was carried previously by A9395, which was also at Harlaxton, UK, while No. 68 (Australian) Squadron was in training. A9242 went to France being recorded in the Squadron diary on 30 September 1917. On 11 October it was extensively damaged in a forced landing at Honnecourt. Returned to the Depot, it was not considered worth repairing and was written off. Note the spanwise aileron connecting cable, early cowl, ring sight, and what appears to be a camera gun mounted on the port side. It is not known whether these markings were carried in combat and photographic proof one way or the other continues to prove elusive. (Credit: via Late N. Hewitt)

B49?? is from a batch of 100 (B4901–B5000) built by March, Jones, and Cribb under contract 87/A/1714. The dark (red) fuselage band and background would indicate a training machine. Note the fabric panel opened to allow access to the carburetor. This machine has the later reinforced cowl.

B377 *The Chiefs of Ashanti No. 3* went to France from Hendon on 24 August 1917. It was one of 14 aircraft of No. 68 (Australian) Squadron when the squadron set up operations at Baizieux on 23 September 1917. Assigned to "C" Flight, it was allotted the identification letter "Z" and usually flown by Lt. Robert W. McKenzie.

Holden landed A9278 at the advanced landing ground with his machine a flying wreck. The aircraft would land at the advanced landing ground, be replenished with fuel and ammunition and repaired, if possible, and then returned to the fray. The weather was so bad that only one German aircraft was seen during the day and this disappeared in the mist. Ward was shot down in A9399 by ground fire and he broke his leg in the ensuing forced landing behind German lines, being taken prisoner. Taylor in A9378 was shot down inside enemy lines. His companion, Wilson, related how the pair had attacked the massed enemy, preventing them from concentrating their fire on the advancing British troops.

Then, as I zoomed up after a burst of machine-gun fire and turned to dive again, I missed Taylor... The next second the red light of a pilot-rocket showed up beside me. I guessed it was fired by Taylor, and it meant that he was in distress. Another red light followed rapidly, and then I saw him down on the ground wrecked and among the enemy. That he was sufficiently alive to fire his rockets was amazing. His machine was just a heap of wreckage. One wing lay twenty yards away from the rest of the heap, from which Taylor had scrambled...

Under protective fire from Wilson, Taylor took up a German rifle, joining up with an advancing British infantry patrol which had lost its officer, and leading it to bring in a wounded man. Wilson's windscreen was hit and he had to pull up to clear the glass dust from his eyes. When he returned there was no sign of Taylor and his gallant little band. Wilson gave him up for dead or captured as the ground was in the possession of the Germans. In the meantime Taylor had fought his way back to the main body of troops. He came across Bell's D.H.5 and tried to fly it out without success. He then rejoined the squadron at the advanced landing ground.

Out of the 18 machines that took part in the attacks, the squadron had lost six shot down, one missing, one pilot missing, one dead of wounds, and one wounded.

On 22 November 1917, Lt. F.G. Huxley in A9461 succeeded in bringing down an Albatros for the Squadron's first victory. That same morning Lt. R.W. Howard (A9294) in company with two S.E.5a scouts drove down a D.F.W. two-seater while Phillips, in A9288, forced another two-seater to land. For their efforts during the Battle of Cambri six members of the squadron received the Military Cross.

The Australians found the D.H.5 outclassed, but with its maneuverability it was able to hold its own when in the hands of a first-class pilot. Captain A. King Cowper (an Australian who flew the D.H.5 with No. 24 Squadron, RFC) recalled that while very maneuverable and a very good fighting aircraft by *having a backward wing stagger it was completely blind for rear vision, therefore a dangerous aircraft; used for ground observation it was ideal, having such a marvelous forward vision.*

The Australians had used their "back-staggers" for less than three months when they were phased out for S.E.5a scouts.

The type was built by Airco, the Darracq Motor Engineering Co., Marsh, Jones and Cribb, and the British Caudron Co. Ltd. It was soon supplanted by aircraft of higher performance.

Specifications:

Span: 25 ft. 8 in. (7.82 m), Length: 22 ft. (6.71 m), Height: 9 ft. 1.5 in. (2.78 m), Wing area: 212.1 sq. ft. (19.70 sq. m).

Engines:	110 hp Le Rhone 9J
	110 hp Clerget 9Z
	100 hp Gnome *Monosoupape*

There was great patriotic endeavor in Australia to raise money to purchase "Battleplanes" and the Australian Air Squadrons' Fund was launched in early 1916 by one C. Alma Barker. A native of New Zealand, Alma Barker became a rubber planter and mine owner in Malaya as well as owning sheep and cattle stations in Australia and New Zealand. He started the Malayan Air Squadrons' Fund in early 1915 followed by the Australian fund. NSW was to denote 30 aircraft and these were numbered by the Australian Air Squadrons' Fund.

It was common to carry the markings of presentation aircraft onto a new machine when the first was lost; however, A9197 bore the inscription for *New South Wales No. 15* and No. 16. It appears that No. 15 should have been D.H.5 A9432 and the number shown was quickly rectified to the correct No. 16. This aircraft must have then been lost as A9245 became No. 16 and went to France with the Squadron.

A9197 was photographed at Harlaxton. It has the early cowl and rubber bungee return springs to the ailerons. Note the fuel line from the gravity tank and the pitot installation on the interplane strut. This aircraft is not recorded in No. 68 (Australian) Squadron's diary. (Credit: G. Muir)

These two photographs from almost identical positions show the different interpretation given to serial marking instructions by the Aircraft manufacturing Co. (A9245) and the Darracq Motor Engineering Co. (A9432). Both aircraft were with No. 68 (Australian) Squadron while it was training at Harlaxton before it went to France in September 1917.

A9432 bears the inscription *New South Wales No. 15*. This aircraft was not recorded in No. 68 (Australian) Squadron's diary along with a number of other D.H.5 fighters which are confirmed as being with the squadron by photographs.

A9245 bore the inscription *"New South Wales No. 16, The Upper Hunter Battleplane; Presented by the residents of Upper Hunter District; New South Wales."* This aircraft became aircraft "C" of "B" Flight when the squadron moved to France. Its pilot was usually Lt. Les Holden. It was recorded as being received on 5 December 1917, however its stay with the squadron was brief as it was returned to 2 AD on 19th of that month when the squadron re-equipped with S.E.5a fighters.

The photograph of A9245 appears to show the fitting of a camera gun which indicates that the aircraft was probably still in the UK at the time. (Credit: via Late N. Hewitt)

The top photograph of Les Holden and A9344 has been credited as being taken after he landed on an advanced landing ground when he was shot up by enemy aircraft during the Battle of Cambri. Holden returned to the advanced landing ground with his elevator control shot away, the bottom longeron, main ribs, left rear undercarriage strut, left and right longeron, tailplanes, petrol tank, and center section struts shot through. In fact everything was shot through except the pilot, who received a bullet in the sole of his rubber boot and another which split his high boots at the knee. Squadron records give A9278 as the aircraft in this memorable fight. This aircraft was sent to the Air Depot on 20 November 1917. A9344, the subject of the photographs, was flown back to the UK on 25 January 1918.

On being received at No. 68 (Australian) Squadron on 24 November, A9344 became Aircraft "C" of "A" Flight. The lower photo appears to show the same location as above and the presence of a woman makes the claim that this was taken on the advanced landing ground unlikely. Note the Aldis sight, strengthened cowl, streamer on rudder, and the cockade on the bottom surface of the top plane. (Credit: L. Holden Collection via late D. Martin)

The pilot's forward view was superb; however, the view to the rear and behind was one cause of the D.H.5's failure as a fighter. The location of the pilot to the Vickers machine gun made access for in-flight stoppages easy. Note the cowl stiffeners. (Credit: A. Revel)

Right: A9435 "E" of No. 24 Squadron, RFC.

Below: A9272 "3" of No. 24 Squadron, RFC.

A946? "6" of No. 24 Squadron, RFC.

A9340 "C" of No. 32 Squadron, RFC.

A9474 of No. 41 Squadron, RFC.

Two views of A9507 of No. 64 Squadron. The individual letter "E" has been used to spell *Elsa*. Capt. E.R. Tempest recorded an "Out of Control" in this aircraft on 30 November 1917. It survived until at least February 1918, when it returned to England. (Credit: A. Revel)

D.H.6

The "Dung Hunter," the "Biscuit Box," the "Clutching Hand," the "Chummy Hearse," the "Flying Coffin," and "Sky Hook" were some of the (printable) names given to Geoffrey de Havilland's attempt to produce a cheap, easily constructed, maintained, and repaired primary trainer.

The needs of the RFC in the Battle of the Somme saw the RNAS loan its men and aircraft to the RFC. The shortfall in the personnel and material of the RFC had to be addressed; the RFC had to expand for the coming critical battles of 1917–1918. To this end the need for a new primary trainer became imperative. The D.H.6 was de Havilland's answer to this need.

Everything was sacrificed for the sake of speed of construction. All major components were straight sided, upper and lower wings were interchangeable, and the resultant aircraft could never be called elegant. The elegant de Havilland fin and rudder of the prototype was soon replaced by a more utilitarian affair. Likewise the rounded top decking of the prototype gave way to a flat structure in the production aircraft.

The airframe was of conventional wooden construction, wire braced and fabric covered with the front fuselage being ply covered for extra strength. The occupants occupied a large communal cockpit and the instructor was provided with a lever with which to disengage the pupil's controls in an emergency.

Power was supplied by a 90 hp R.A.F. 1a V-8 air-cooled engine attached directly to the top longerons without any cowling other than a large air scoop. As production outstripped supplies of R.A.F. engines many were completed with 80 hp Renault or 90 hp Curtiss OX-5 engines. The Curtiss engine was enclosed with a frontal radiator.

Production began in January 1917 and the type was widely used in the UK, some even finding their way to Australia and it was proposed to build the type in Canada. It was not considered a good trainer as its performance was little better than the Farman Shorthorns already in service. However, it had no vices and would remain airborne at an air speed of 30 mph (48 kph) and was considered too safe even for a primary trainer. Its lack of speed is illustrated by the story of a Russian pilot who taxied to the airfield boundary on a particularly windy day and took off into wind and proceeded to travel backwards across the airfield, landing at the other boundary!

The type was replaced by the Avro 504 in late 1917 for training but continued to serve in the anti-U-boat patrol squadrons. Losses were particularly heavy on the north-east coast and in the south-west as ships headed for ports. A submarine could not tell that the aircraft was a D.H.6 or of more warlike mien and would submerge whenever an aircraft was sighted. Initially proposed as a "temporary expedient," 34 D.H.6 flights were to be formed, operating from various coastal stations until the Armistice. It was proposed to have a series of "protected lanes" which would be patrolled by an aircraft every 20 minutes. Performance was so poor that when carrying a bomb load an observer could not be carried. About 75% operated in this way. Ground conditions at theses flights were poor and engine failures were common. Ditchings were frequent. The unexpected

Prototype D.H.6, note typical de Havilland fin and rudder. (Credit late Sir L. Hartnett.)

capacity of the aircraft to stay afloat for as much as six hours saved many lives. The D.H.6 crews were in frequent contact with submarines.

In an attempt to improve performance new wings with reduced undercamber were fitted and rigged with 10 inches (305 mm) of back stagger. A smaller rudder and elevators were fitted. Speed was improved slightly but this was all.

The US Naval Aviation Repair Base, Eastleigh, England, listed, on 31 January 1919, the following ten D.H.6 aircraft "procured from (the) British Government": C7736, C7741, C7745, C7747, C7753, C7754, C7756, C7757, C7759, and C7760.

Unusually for such an aeroplane, the D.H.6 had a long civil career in Australia, South Africa, and New Zealand. In the USA, remodeled D.H.6 biplanes having forward stagger and individual cockpits were offered for sale. The Spanish government used a number (stated as 45 to 60 in various sources) in its air force.

The D.H.6 was remembered with affection by one writer as
God's gift to the trainee… the dearest, kindest, most sedate old lady in a first solo. You could stagger "off the deck" with her, do the flattest of turns, and then just fall back on to the ground, and she would still allow you to remain in one piece.

Specifications:

Span: 35 ft. 11 1/8 in. (10.95 m), Length: 27 ft. 3.5 in. (8.32 m), Height: 10 ft. 9.5 in. (3.29 m), Wing Area: (D.H.6) 436.25 sq. ft. (40.53 sq. m) (Modified with back stagger) 413 sq. ft. (38.37 sq. m).

Engines:	90 hp R.A.F. 1a
	80 hp Renault
	90 hp Curtiss OX-5

A D.H.6 photographed at No. 20 Training Squadron, Harlaxton. (Credit: G.N. Moore)

The use of the D.H.6 for anti-submarine patrols led to experiments with flotation bags being conducted at the Isle of Grain. This photograph dated August 1918 shows the air bags in the stowed position.

Above: Wreck of D.H.6 B??62 which flew into the hangar at No. 20 Training Squadron at Harlaxton. Note the fuselage stripe and band behind the cockade. (Credit: G. Moore collection)

Left: The low speeds and light construction of 1914–1918 aircraft led to some spectacular non-fatal crashes. This D.H.6 has chosen an unlikely stopping place. The white triangle is possibly the marking of No. 13 Training Squadron, Yatesbury. Note the "7" on the fin.

C9373 was one of two acquired by the Pratt brothers from the RAF in Egypt and shipped to Australia. The other was C9372. Arriving in December 1919, they survived the hectic pre-registration days to become G-AUEA and G-AUDO respectively. Note lifting markings on lower wing tip.

A9652 was unusual in that the last digit of the serial is in white on the red rudder stripe. This was probably a replacement rudder as no manufacturer used this style of marking. (Credit David McGuiness collection)

Eight D.H.6 trainers, the "tuition machine of the future", were ordered on 20 March 1918 for the Australian Flying Corps Central Flying School (CFS) at Point Cook, Victoria. Two were lost at sea, and in the meantime the first had been erected and the results must have been disappointing as the Australian authorities cancelled the order to replace the missing two aircraft. These had already been dispatched and so eight were operated by the CFS. Known serial are B2801(?), B2802, B2803, B2804, C9372, and C9374. Six of the CFS D.H.6 trainers were later sold to civilian buyers. Note the lower wing in the background with part of the legend "War Loan" from the 1918 flight promoting the £20 million Liberty Loan. The airframe bears the CFS serial "62". Maurice Farman CFS 15 in the background indicates that photograph was taken at Point Cook. (Credit: RAAF Museum)

Above: A9689 on a snow covered airfield. It was one of a batch of 200 (A9563 to A9762) delivered by the Graham-White Co. Note the replacement aileron on the port upper wing. Grahame-White's difficulties with the timber forced upon his company to build the D.H.6 are detailed in his biography (*Claude Grahame-White*, Wallace G. Putman, UK, 1960). His protests that the timber was not up to strength were ignored and the aircraft were condemned. Grahame-White eventually had to sue the government to gain recompense for the cost to his company of the condemned material.

Poor but interesting photograph of a D.H.6 trainer with cockades on the upper surfaces of both wings and a broad white fuselage band. (Credit: via N. Hayes)

C9341 at Biggin Hill in September 1918. The checker board panel may identify it as belonging to the Wireless Testing Flight. (Credit: David McGuiness collection)

A civil D.H.6 racing a Sopwith Swallow, probably around Melbourne, circa 1921.

Twin gravity tanks have been fitted to this civil Australian D.H.6. The start of the registration G-AU?? can just be made out under the lower wing.

Spanish D.H.6 with 180 hp Hispano Suiza engine. Note the Lamblin radiators, airfoil gravity tank center section, and typical de Havilland fin and rudder. Sources vary on the number obtained by Spain; from 45 so modified in 1920 to 60 built under license at Guadalajara from 1921. A more likely figure is two civil imports and 10 constructed for the *Aviación Militar* circa 1924. They were used at the main training establishment at Cuatros Vientos. A few may have ended up on the civil register. (Credit: via J. Miranda)

This view shows the utilitarian lines of the D.H.6. Note the cockade on the bottom surface of the upper wing. Photograph probably taken at Point Cook by a very youthful Sir Wilfred Brooks. (Credit: Sir W. Brooks)

Above: Aerial advertising carried to the fullest! Every surface except the top surface of the upper wing has been utilized for advertising on this ex-CFS D.H.6 biplane. Note the cockades on the upper wing.

Right: G-AUBH of Shaw-Ross Aviation over Melbourne. Formerly B3204, it was imported in June 1921.

D.H.9

The success of the D.H.4 was over by the summer of 1917. An aircraft capable of carrying heavier loads over longer distances was required. The Air Board was reluctant to give up the infrastructure which had been developed around the production of the D.H.4 and so it was not surprising that they agreed to the large scale production of the D.H.9 which made considerable use of existing standard D.H.4 components. The mainplanes and tailplane were the same. The pilot and gunner had been moved closer together. The engine was cowled in a "Hunnish" manner with a retractable radiator mounted under the fuselage. The prototype was a modified D.H.4, A7559, fitted with a 230 hp Galloway-built B.H.P. Adriatic engine. The first flight of the prototype was in July 1917, and D.H.4 contracts were converted to D.H.9 bombers.

Initially fitted with the Siddeley-Deasy Car Co.-built B.H.P., the majority of aircraft were fitted with the Siddeley Puma, a light-weight version of the B.H.P. modified for mass production. The Puma had considerable problems and instead of delivering the expected 300 hp, it was de-rated to 230 hp. As a result the D.H.9 was under-powered and inferior to the D.H.4, the aircraft it was to replace!

With a military load comprising 4.5 gallons (20.5 liter) of oil, 70 gallons (318 liters) of fuel, 6.5 gallons (295 liters) of water, four 112 lb. (51 kg) or two 230 lb. (104 kg) bombs,

D.H.9 C6065 was a training aircraft in the UK when this photograph was taken. Note the camera gun on the observer's Scarff ring and lack of armament for the pilot. The aircraft is doped PC 10 over all upper surfaces. It is known to have been at the School of Navigation and Bomb Dropping by 20 March 1918. (Credit: Late F.S. Briggs)

Another D.H.9 training machine. Photograph probably taken after the Armistice. Note the large container to catch spent cartridges, a post-war austerity measure, and the under fuselage bomb. (Credit: Bargwana collection via N. Hayes)

Built by the Aircraft Manufacturing Co. as part of a batch of 200 (E8857 to E9056), E9029 has obviously had a mishap! It was with No. 206 Squadron when this photograph was taken. The fuselage metal and ply panels are gray while fabric covered surfaces were PC 10 with clear doped lower surfaces.

Vickers gun for the pilot, and Lewis gun for the observer, it was unable to climb above 13,000 feet (3,960 m). As well as this, the Puma gave considerable trouble and there were many engine failures. This put an intolerable burden on aircrew. It was essential for aircraft in daylight bombing operations to keep formation, the key to mutual defense. In August 1918 Trenchard's insistence led to the withdrawal of the type from operations on the Western Front, but this took time and it was still on operations at the time of the Armistice. The D.H.9 did not replace the D.H.4 but supplemented it. In areas where the Allies had aerial superiority, such as Palestine, the type performed creditably. Even here forced landings were a hazard.

The problems faced by the D.H.9 are illustrated by No. 99 Squadron's experiences of 31 July 1918. The formation leader decided that it was impossible to reach the primary target of Mainz due to the intensity of enemy opposition and so diverted to Saarbrücken. Before the new target could be reached four D.H.9 bombers had been shot down. Three more were lost in the running fight on the homeward run. Only two aircraft returned across the lines. The Squadron lost 14 airmen (four killed, one died of wounds, and nine as POWs) and it would be some time before replacements could be trained in formation flying and the squadron become active again. The Squadron's history noted that the "D.H.9 was an extremely vulnerable machine." Similarly, No. 104 Squadron had to stop operations and recover three times after losses.

E9001, in standard camouflage and equipment, was from the same batch and also served with No. 206 Squadron. Lt. C. Workman and Air Mechanic E. Rogan became lost in fog and force landed in the Netherlands on 3 March 1919, both being injured in the resultant crash. It is presumed to have been written off at this time.

Not everyone thought badly of the Puma engined D.H.9. Sqn. Ldr. A.H. Curtiss has recorded that *of all my D.H.9 flights, totaling approximately 180 hr, about 13% were marred by engine troubles; which considering their almost daily usage, was not a bad record.*

Curtiss flew D.H.9 bombers in No. 49 Squadron, RFC/RAF. On 7 June, 1918, he took part in a raid on "hunland" wherein he experienced the vagaries of the Puma engine while flying C6114, aircraft "M." After bombing their target with ease the formation was bracketed by anti-aircraft fire. The Germans were alerting their fighters to the presence of the British force. Curtiss tried counting the scouts in the German circus which came in answer to this call but gave up after he reached 30. Apparently aircraft from *Jastas* 15 and 22 were involved in the ensuing combat.

Lt. G.C. McEwan and Lt. T.E. Harvey were at the back of the "Vee" formation of seven, in the opposite position to Curtiss. Their machine, C6184, began to pour smoke and soon turned over to disappear against the ground below. Curtiss tried to be clever by not making a symmetrical (evasion) pattern. *Of all things my engine chose to start packing up soon afterwards. I tried everything, going over from main to gravity and back, testing magneto and coil switches and generally attempting to put things right; all to no avail.*

Diving well beyond the "never exceed speed", Curtiss looked back to see his tail waving in the slipstream in such a way to indicate that two bracing wires had been shot away. He eased out of the dive before the tail disintegrated. His observer on this occasion was Sgt. A.W. Davies. (His regular observer, P.T. Holligan, was due to go on leave and an unwritten law in the squadron was that such a person did not fly on the previous day.) Davies had his Lewis gun shot and put out of action and in an act of *sheer Bravado, he fired his Very pistol over the circus without much chance of scoring. The huns must have misinterpreted this as a call for help, with the result that the entire fleet turned tail.*

Curtiss landed on the Allied side of the front lines "but only just." The D.H.9 made an ordinary if not graceful landing. The crew discovered that the aileron spanwise compensating cable had snapped. The first anti-aircraft burst had hit the radiator which had a hole big enough to put a fist through and the engine had seized solid. For the loss of two aircraft, as C6114 was written off, the squadron claimed five German fighters. All in all, Curtiss considered this a good result.

The type served on the Western Front with Nos. 17, 27, 47, 49, 98, 99, 103, 104, 107, 108, 144, 202, 206, 211, 212, 218 to 224, 226, 233, 236, 250, 254, 269, 273 Squadrons and the 17th Wing operating from Malta. Belgium acquired 18 in 1918. The AEF acquired two airframes in July 1918 to try with the 435 hp Liberty 12A engine in an attempt to repeat the success of the Liberty D.H.4.

The end of the war saw the D.H.9 continue in RAF service in Russia with Nos. 47 and 221 Squadrons, and the Middle East. D3117 was modified in Somaliland in 1919 to carry a stretcher case in a large enclosure built on top of the fuselage. South Africa, New Zealand, Canada, and Australia received the type as part of the Imperial Gift. Afghanistan, Bolivia, Chile, Estonia, Greece, the Irish Free State, Latvia, the Netherlands, Poland, Soviet Russia, Spain, and Switzerland acquired the type. Some survived in Spanish service to participate in the Civil War. At least one, Serial 34-18, survived until 1940. In peace time it achieved some credible firsts but the Puma was always its "Achilles' heel."

The D.H.9 went on to have a successful civil career. It was

Aircraft "A" of "B" Flight, No. 211 Squadron 1 April 1918. The observer is Lt. W. Norrey. Note how the "A" has been used to spell 'Acme', the significance of which is not known. (Credit: W/Cmdr. Norrey)

modified to have a cabin and many were used in the United Kingdom, Australia, Denmark, Spain, New Zealand, and Belgium. The ultimate civil version was the D.H.9J, which featured a strengthened front fuselage to carry an 350 hp Armstrong Siddeley Jaguar III engine. These were in service from 1926 to 1936.

Specifications:

Span: 42 ft. 4 5/8in. (14.21 m), Length: 30 ft. 6 in. (9.30 m), Height: 11 ft. 2 in. (3.40 m), Wing area 434 sq. ft. (40.32 sq. m).

Engines:
        230 Siddeley Puma
        230 hp Galloway Adriatic
        260 hp Fiat A.12

D.H.9 aircraft "N" of No. 49 Squadron in German hands. It has force landed on a beach. (Credit: P.M. Grosz)

Above right, right, and below: In the Middle East C6293 of No. 114 Squadron bears an unusual color scheme.

D.H.9 "M" of No. 49 Squadron, RAF.

D.H.9 "Z" of No. 49 Squadron, RAF.

D.H.9 "N" of No. 49 Squadron, RAF, in German hands.

Production of the D.H.9 was prodigious, some 3,890 being produced out of orders for 5,534. The first aircraft in this line up is D1269, which indicates that this photograph could have been taken at National Aircraft Factory No. 2, which produced D1001–D1500.

The Netherlands interned at least 15 D.H.9 bombers during the 1914–1918 War. These aircraft were documented as they were acquired and give a good overview of operational D.H.9 bombers in late 1918. Ten were given *Luchtvaartafdeling* (LVA) serials: deH433, deH434, deH437 to deH439, deH441 to deH444, and deH446. This did not mean that they were flown, as serials were issued to aircraft which appeared to be repairable but were never repaired. B7620 "A" of No. 211 Squadron was the first thus acquired on 27 June 1918. The basic PC10 camouflage scheme has been liberally overpainted with white stripes and included eyes and a mouth under the nose cone. Note the exhaust pipe. During an attack on Brugge, Capt. J.A. Gray and F/Lt. J.J. Comerford were involved in an air fight and then hit by AA fire. Over Holland they were under continuous fire, finally landing when their engine stopped. At first this aircraft was not purchased but later it was by the Colonial Office and used for training pilots at Soesterberg for the Netherlands East Indies Army's air corps, the *Koninklijk Nederlandsch Indisch Leger* (KNIL). (Credit: F. Gerdessen)

Unknown D.H.9 in Italy. The marking appears to be the numeral "1." (Credit: R. Gentilli)

Camera gun photograph of a D.H.9. (Credit: Late Hasting-Derring)

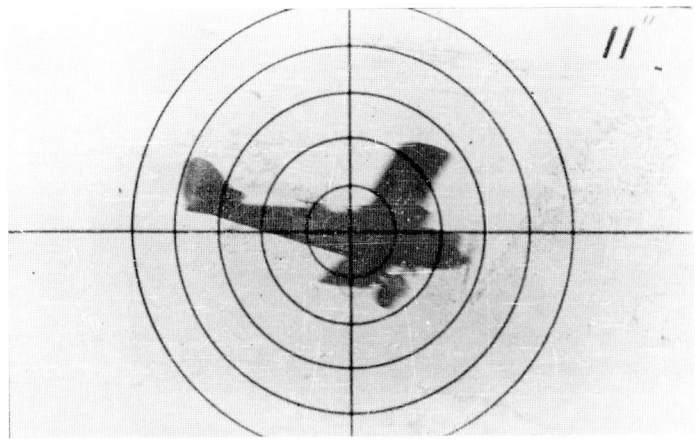

D.H.9 dismantled for road transportation. Note the replacement aileron on the starboard lower mainplane and the aircraft numeral "1." (Credit: Late Hasting-Derring)

The Puma engine was, according to AVM R. Williams, RAAF, the most unreliable engine he ever came across, his experiences in Palestine coloring his view. This D.H.9, serial C6297, of No. 144 Squadron has force landed, wiping off its undercarriage in the process. The crew placed the white ring to indicate that a landing should not be attempted. Note the individual letter "3" and white fuselage stripe.

D1708 "VI" of No. 218 Squadron. Lt. A.C. Lloyd and 2/Lt. M.G. Wilson were apparently credited to *Oblt*. Theo Osterkamp of *Marine Feld Jagdstaffel* 1. The machine was wrecked in landing at Groede, Zeeland. Note that the aircraft individual numeral "VI" is repeated on the top decking. (Credit F. Gerdessen)

Another of No. 211 Squadron's D.H.9 aircraft was D7204 "J" which arrived on 24 August. The pilot, Lt. J.A. Dear, was an American, and the observer was a Scotsman, 2/Lt. J.F.J. Peters. Hit by AA fire over Brugge, they landed near a farm at Zuidzande. Note the elaborate "J" and wheel cover markings. (Credit: F. Gerdessen)

The crew of D3107 "P" of No. 108 Squadron, 2/Lt. J.B. Cox and Lt. J.J. Lister, was fortunate to survive to be interned. On 15 September 1918, the Squadron met a large force of enemy fighters and three D.H.9 bombers were credited as shot down in the raid on Bruges, three making forced landings on reaching the lines and another crashing on landing. All three reported as shot down were interned in the Netherlands. Cox landed at Souburg and ran into a canal. (Note that the Dutch report that Cox was the observer and that he was wounded in the left shoulder.) The machine fell while being lifted from the canal and killed a soldier. It was thus damaged beyond repair. (Credit: F. Gerdessen)

Aircraft "G" of No. 108 Squadron was D1733. It was apparently shot up on the same raid and landed in the Netherlands. Despite the damage it suffered on landing it was given the Dutch registration deH443. The crew of Capt. W.R.E. Harrison and 2/Lt. C. Thomas were interned. (Credit: F. Gerdessen)

Below: The British tried to sell the Netherlands the D.H.9 bombers they had interned; however, they were considered too old and unfit for service. Sixteen British interned aircraft were collected and handed back to Capt. Manning (RAF) at Schiphol on 20 March 1920, subsequently being destroyed there. There may have been nine D.H.9 bombers amongst these. The Netherlands purchased new machines post-war. H-122 of the KNIL shows the cat insignia of the 1st *Vliegtuigafdeling* on the upper rudder stripe. Note the opening in the top fuselage decking behind the rear cockpit. (Credit: F. Gerdessen)

Close-up views of the D.H.9 cockpit area. Note the Vickers gun mounting, the ring sight, and twin wind driven pumps which kept fuel up to the gravity tank mounted in the wing center section. (Credit: Bragwanna collection via N. Hayes; Late B.P. Flanagan; Late Hasting-Derring)

Below and below right: KNIL D.H.9 biplanes in flight. (Credit: F. Gerdessen)

A version of the D.H.9 known as the HL was built by the KNIL workshops. It had a flat frontal radiator and a fully cowled Puma, full plywood fuselage, and horn balanced ailerons and elevators. Some of the standard D.H.9 bombers were also fitted with the enlarged elevators and ailerons. As late as 1934 remaining KNIL D.H.9 biplanes were rejuvenated with Wright Whirlwind engines as illustrated here. (Credit: F. Gerdessen)

The RAAF received 28 D.H.9 biplanes as part of the Imperial Gift aircraft. Given the RAAF serial prefix "A6" the type did not have a good record in RAAF service. They were used for army cooperation as there was nothing else to use, and for training. It is here that most accidents occurred. At least 14 were written off after crashes or flying accidents for a total of eight fatalities. A6-2? illustrated is in PC 10 overall with clear doped lower fabric surfaces. The RAAF modification to the exhaust pipe to take the exhaust fumes away from the cockpit should be noted. The aircraftman standing on the bracing wire for support is the reason that a footboard was introduced for the D.H.9 and D.H.9A in RAAF service.

RAAF D.H.9 biplanes were later doped V84 aluminum overall as shown here. The nearest aircraft, A6-4 (formerly D1129), was stored until it was reconditioned by the Shaw Ross Engineering and Aviation Co. early in 1924. It was then leased to the Civil Aviation Branch of the Defence Department as G-AUEN and operated by Larkin Aircraft Supply Co. for air mail flights until the aircraft on order from the UK could be delivered. Returned to the RAAF on 18 December 1924, it went to 1 Flying Training School (FTS). It was reconditioned by Pratt Brothers in 1926 and operated by 1 Squadron with Puma 83588. In March this engine was replaced due to a leaking water jacket, a common fault with the Puma engine. The replacement Puma 8341/55740 did not fare much better, being replaced in September after 81.10 hours for the same reason. By September 1927 it was with 3 Squadron based at Richmond, NSW. On 14 April 1928, P/Off. Frank C. Ellsworth made a forced landing at Mascot due to a defective radiator. Whilst taxiing D.H.60 G-AUPP landed and ran into A6-4 at right angles. Damage must have been minimal as it was back in service on the 18th when it reported further engine trouble. On 11 June, again at Mascot, Sgt. R.F. Somerville and LAC P. Norris suffered a sudden shock when the starboard shock absorber collapsed, the machine standing on its nose. By November it was at 1 FTS. In December enquiries were made about a broken longeron and it was still unserviceable at 1 FTS in February 1929. It appears it was never repaired as it was approved for destruction in May and reported destroyed in August.

The second aircraft in the line-up with the front cowl detached is A6-15 (formerly D3189). This aircraft was apparently stored until circa June 1923. In 1924 it went to 3 Squadron at Richmond, NSW, where the photograph was probably taken. In February 1927, Puma 55477 was removed after 0.40 hours due to the water jacket leaking. In April Puma 55477 was removed for the same reason after 11.20 hrs. July saw water leaks in both blocks of Puma 82892 after 32.55 hrs. March 1928 saw Puma 82892 suffer a cracked cylinder head after 18.00 hrs. And so the history continues with F/Off. Clive W. Lord with P/Off. Donald McK Carroll suffering a forced landing from 4,000 ft (1,220 m) due to the engine cylinder head blowing out of Puma 83531. On 30 April 1928, F/Off. Clive W. Lord with AC1 Finn as passenger, had a major crash when he stalled on take off from Goulburn, NSW. The evidence suggested that tail heaviness was the cause of the crash and tests were arranged with D.H.9 biplanes loaded with "Overland Kit." Approval for conversion was granted on 6 June.

Remains of a RAAF D.H.9. The D.H.9 was the only Australian Imperial Gift aircraft to enter the civil register, two being leased for air mail work in 1924 and three being given to aero clubs when the RAAF terminated their service in 1929. Most were destroyed. (Credit: RAAF)

Crashed Soviet D.H.9. This aircraft is probably finished in the green over pale blue camouflage scheme used by the Soviets in the immediate post-Revolution years.

The D.H.9 served with the RAF in Russia with 47 and 221 Squadrons. Some were supplied to the Russian government and as can be seen, a number ended up in Soviet service. Note the Nieuport and what appears to be an R.E.8 in the background.

South Africa received 48 D.H.9 biplanes as part of the Imperial Gift and another one was presented by the City of Birmingham. They were given South African Air Force (SAAF) serials 101 to 119 in apparently random sequence. The type gave sterling service to the SAAF going into action during the Rand revolt of 1922, and against the Bondelzwart Hottentots from May to July the same year. The type was rebuilt with the Jupiter VI as the M'pala I general purpose type or with the Jupiter VIII and oleo wide track undercarriage as the M'pala II. M'pala I Serial 151 is illustrated. The last seven were declared obsolete in 1937 and sold to Aero Clubs for a nominal £10. Two survived until the 1939–1945 War, becoming SAAF 2001 and 2005. They were used only as ground instruction airframes receiving Instructional Serials IS-7 and IS-8 respectively in 1942. IS-8 was struck off charge on 19 April 1943 but was not destroyed and is preserved today in South Africa. (Credit: J.M. Bruce/G.S. Leslie collection)

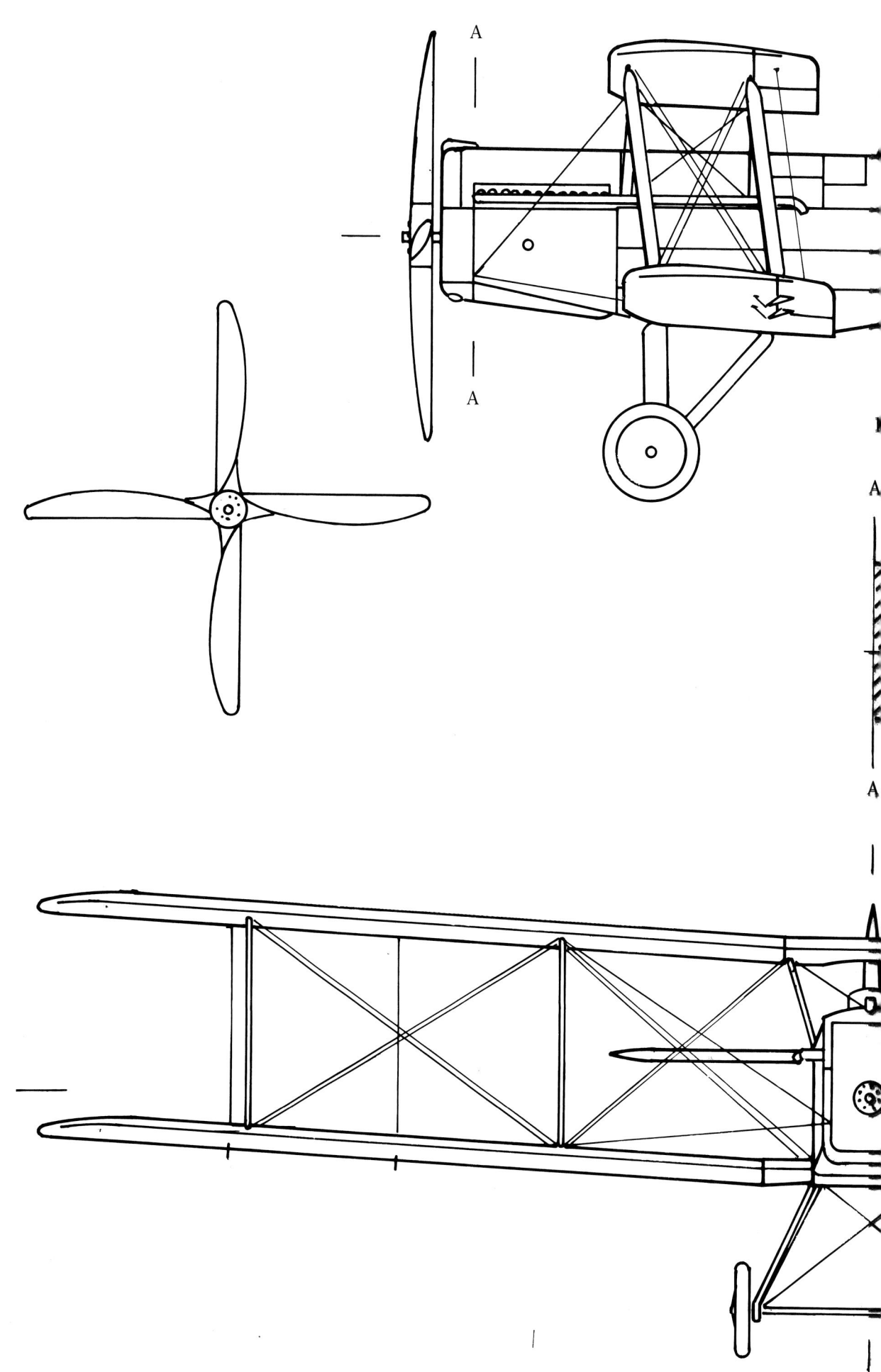

.14

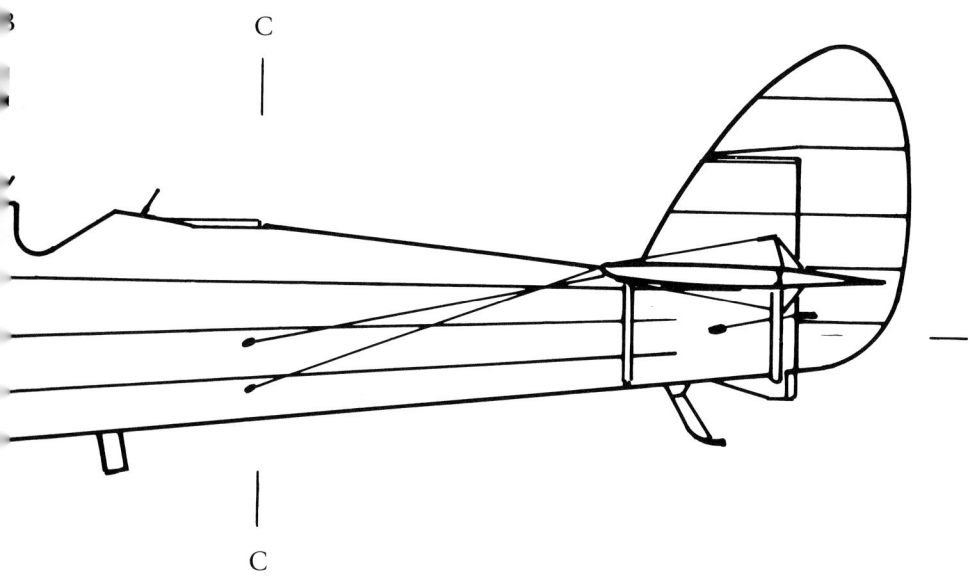

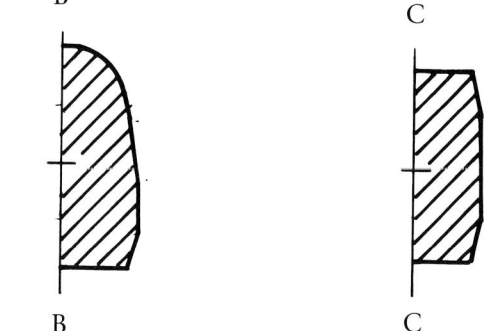

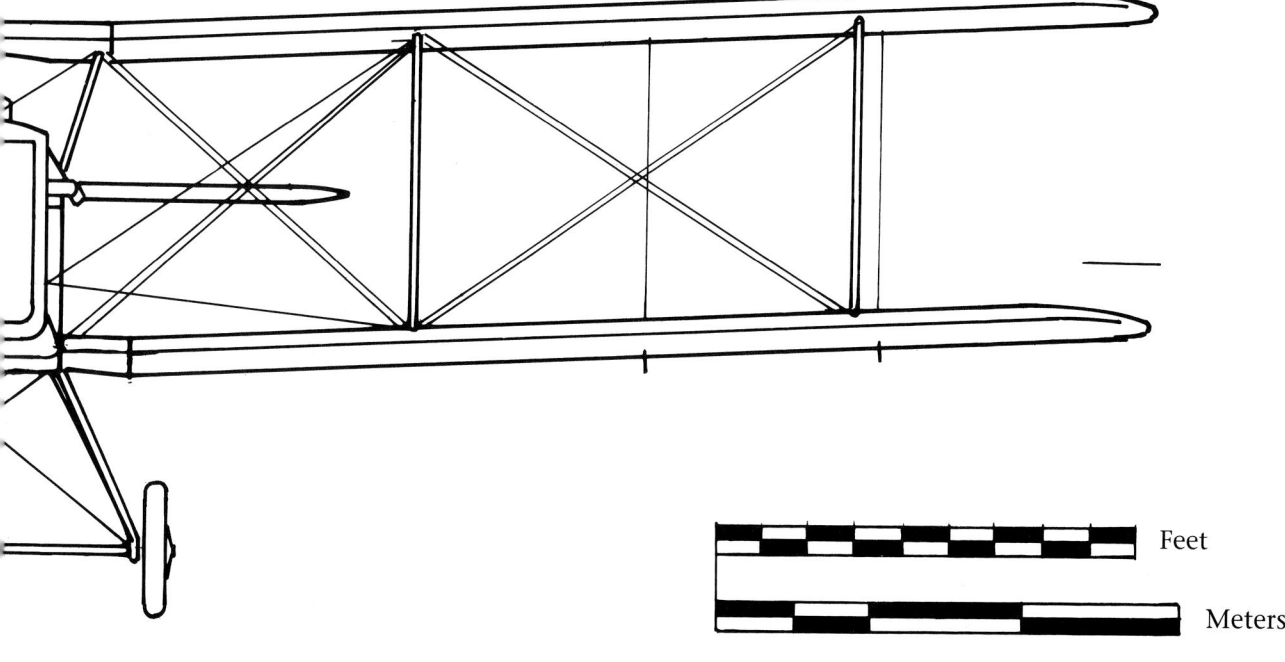

Feet

Meters

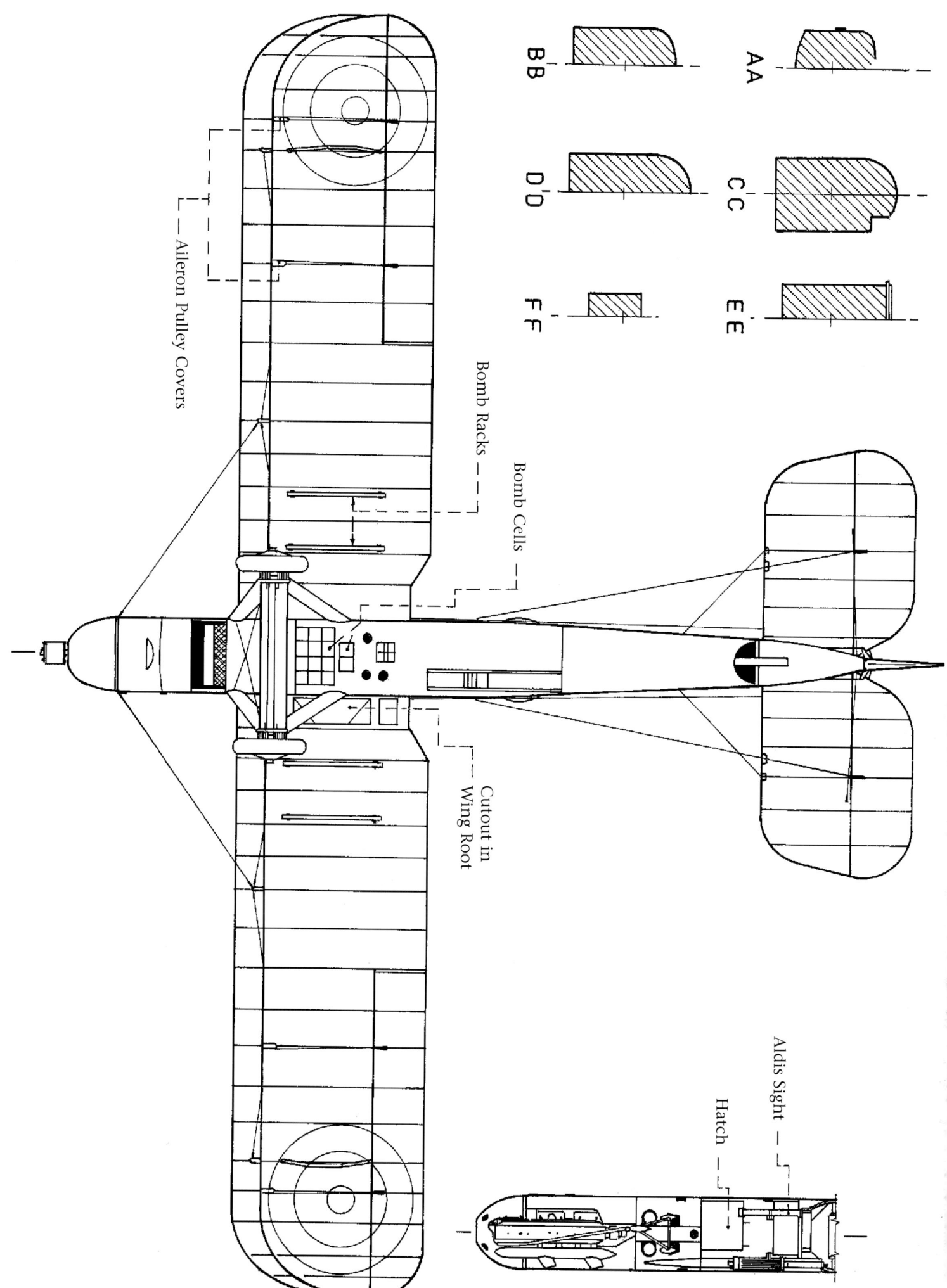

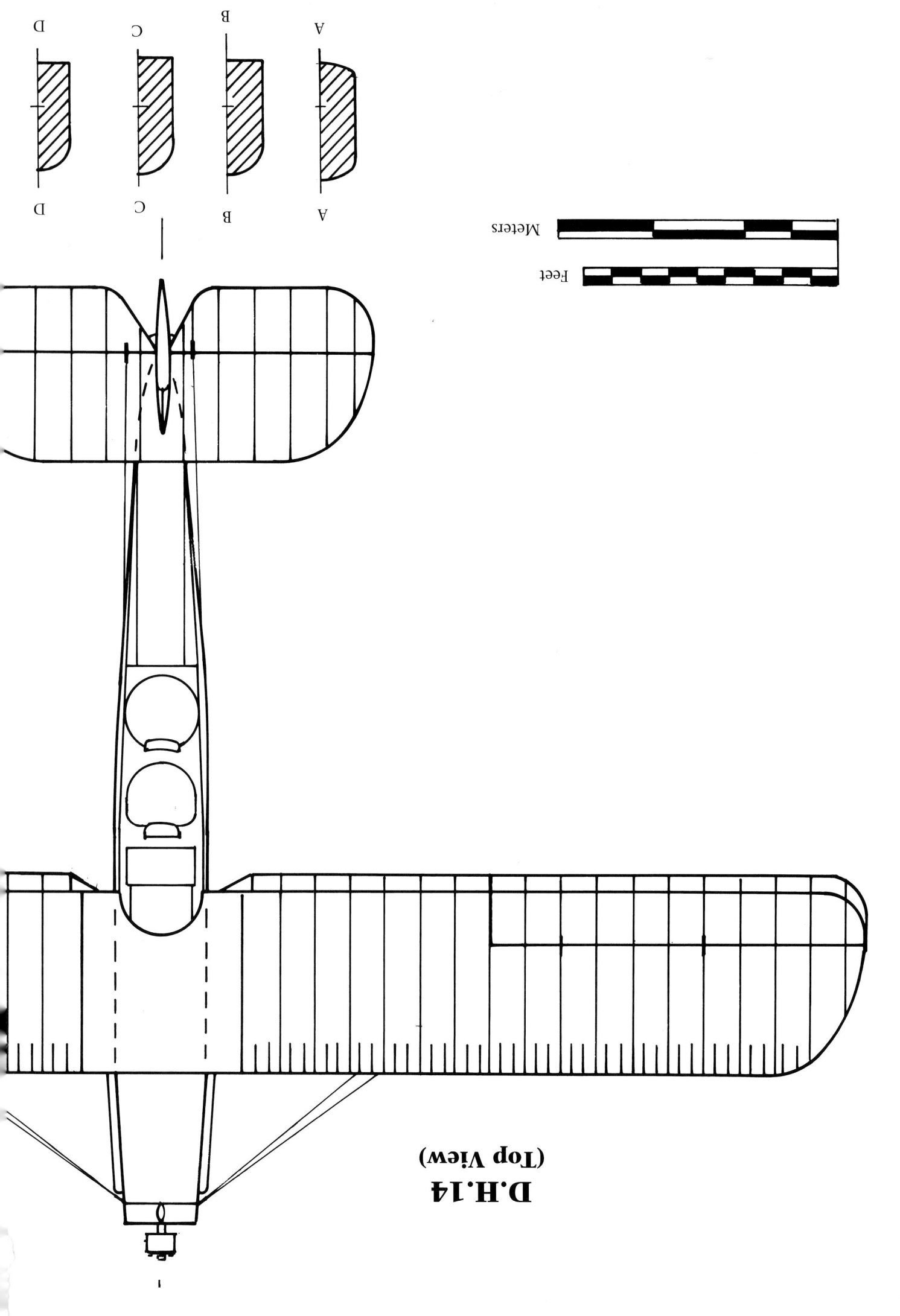

D.H.9A
(Side Views & Fuselage Sections)

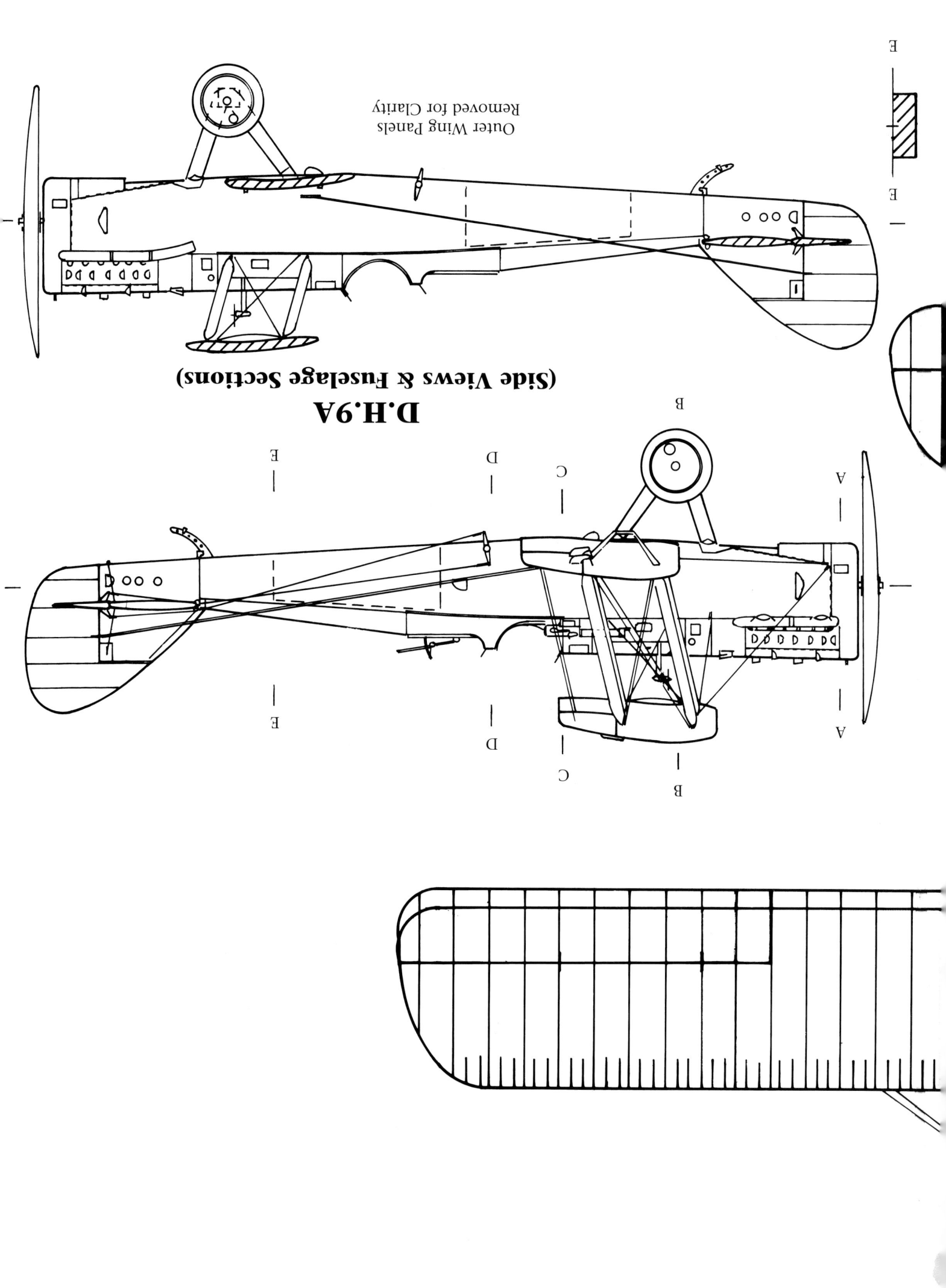

Outer Wing Panels Removed for Clarity

Detail Showing Exhaust Variation

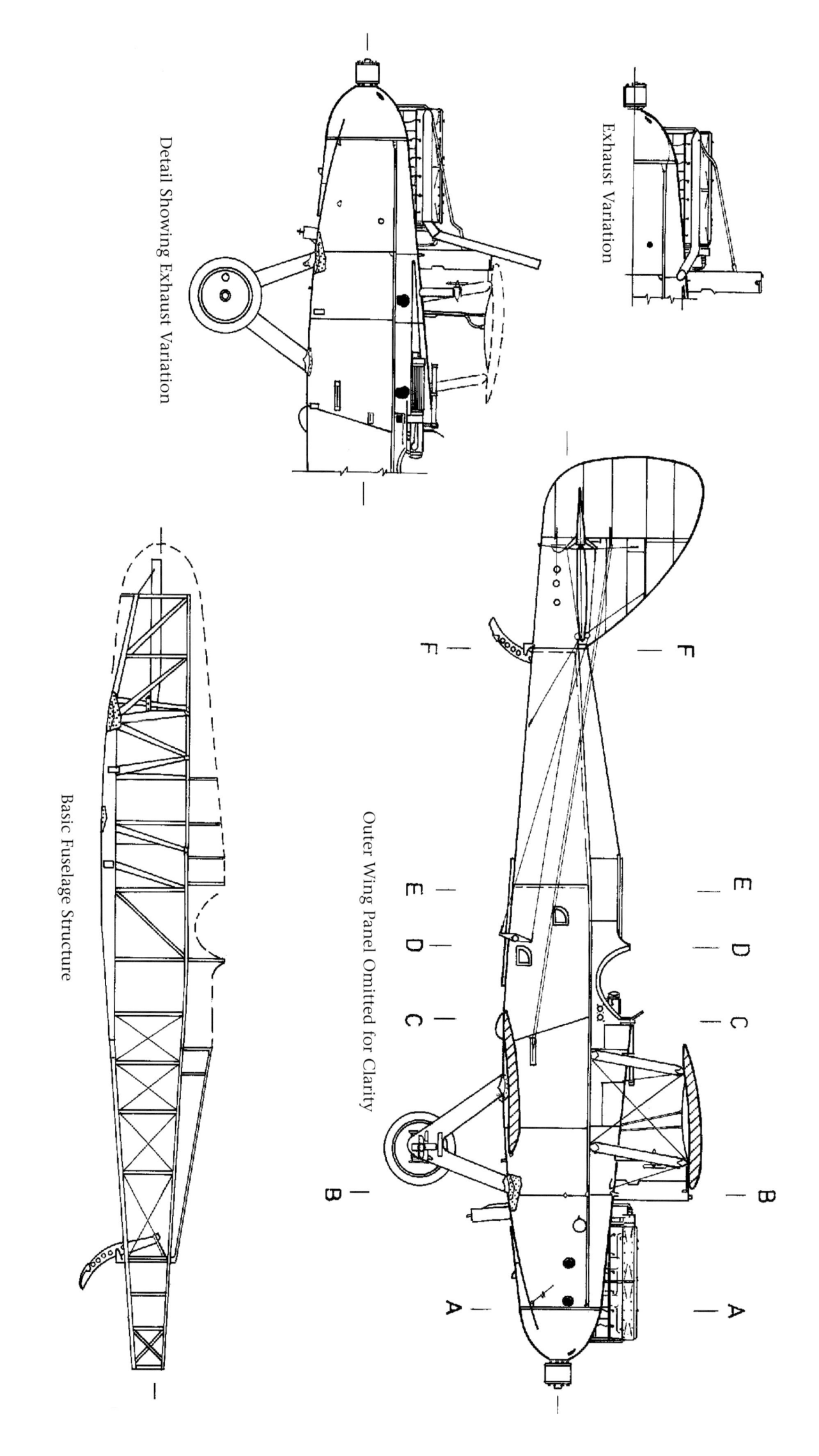

Exhaust Variation

Basic Fuselage Structure

Outer Wing Panel Omitted for Clarity

D.H

.10

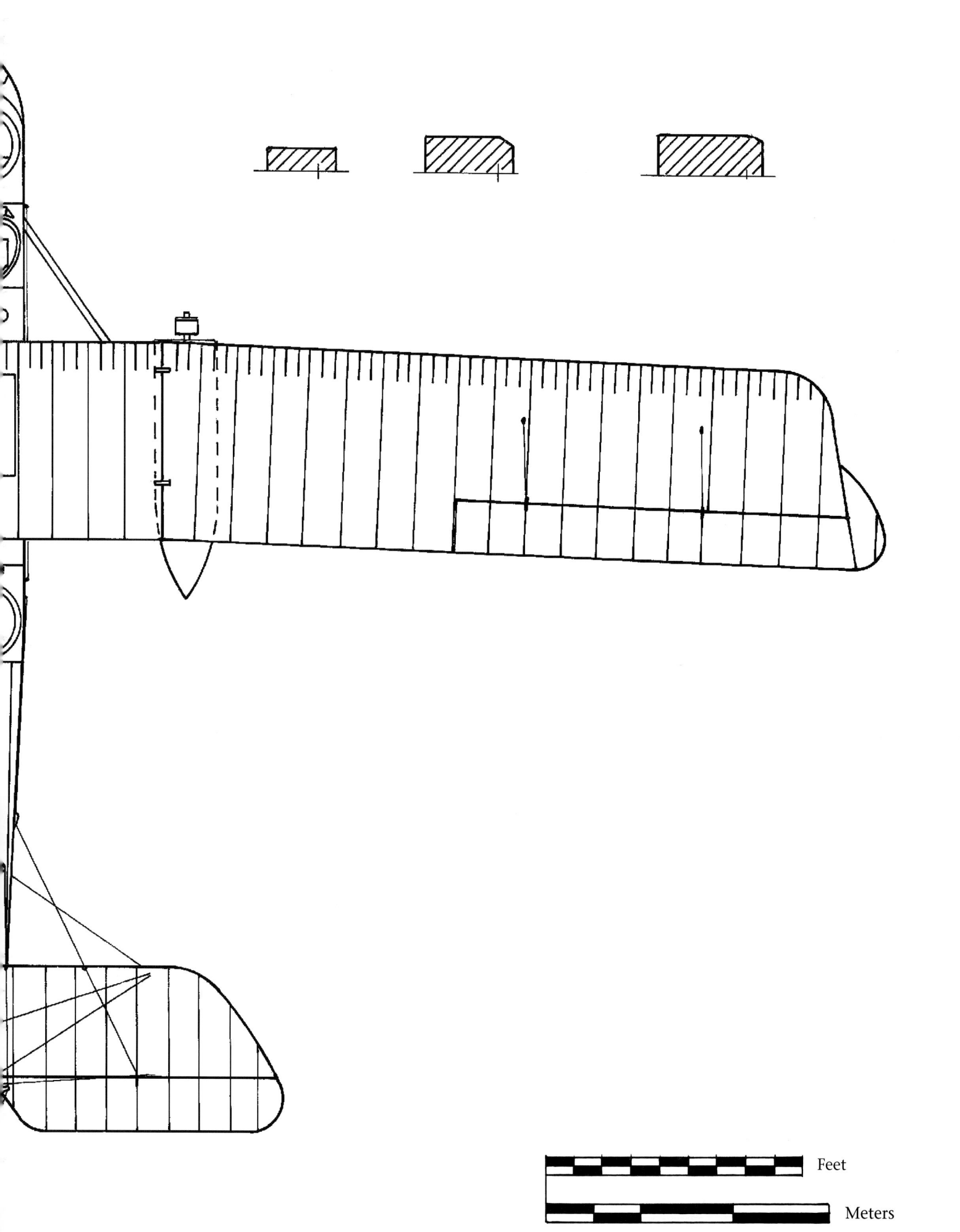

Feet

Meters

D.

Cutout in Wing Root

Feet
Meters

H.9

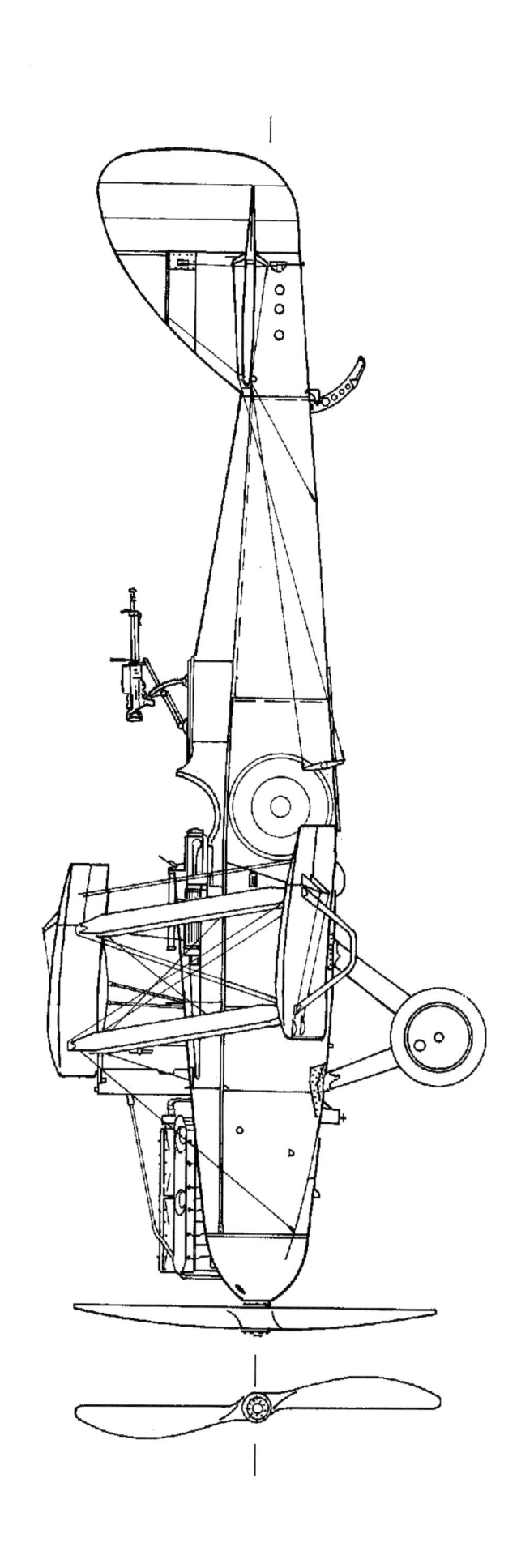

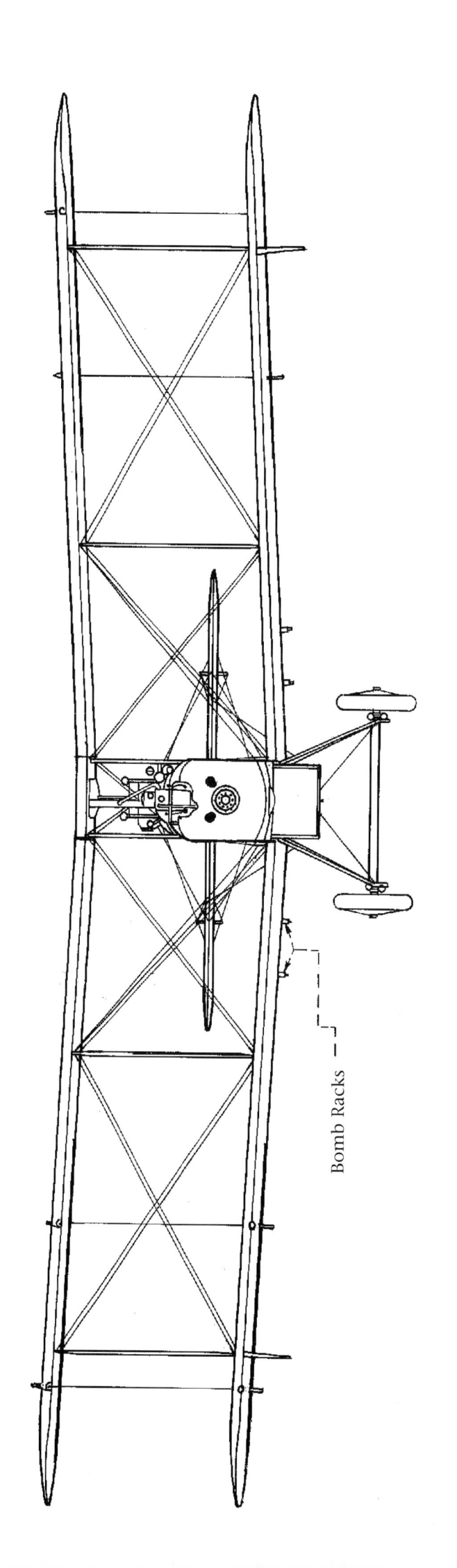

D.H

D.H.11

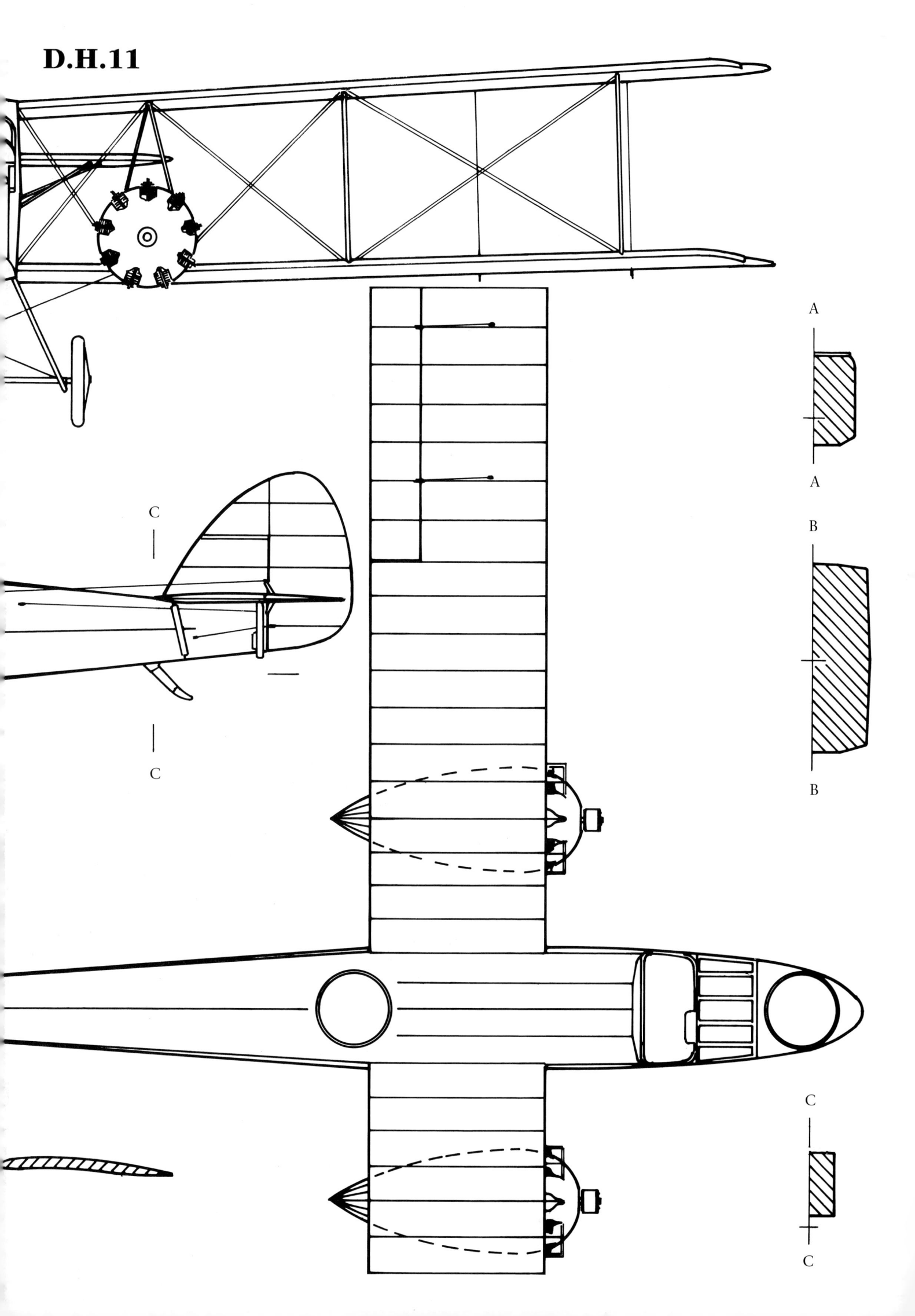

D.

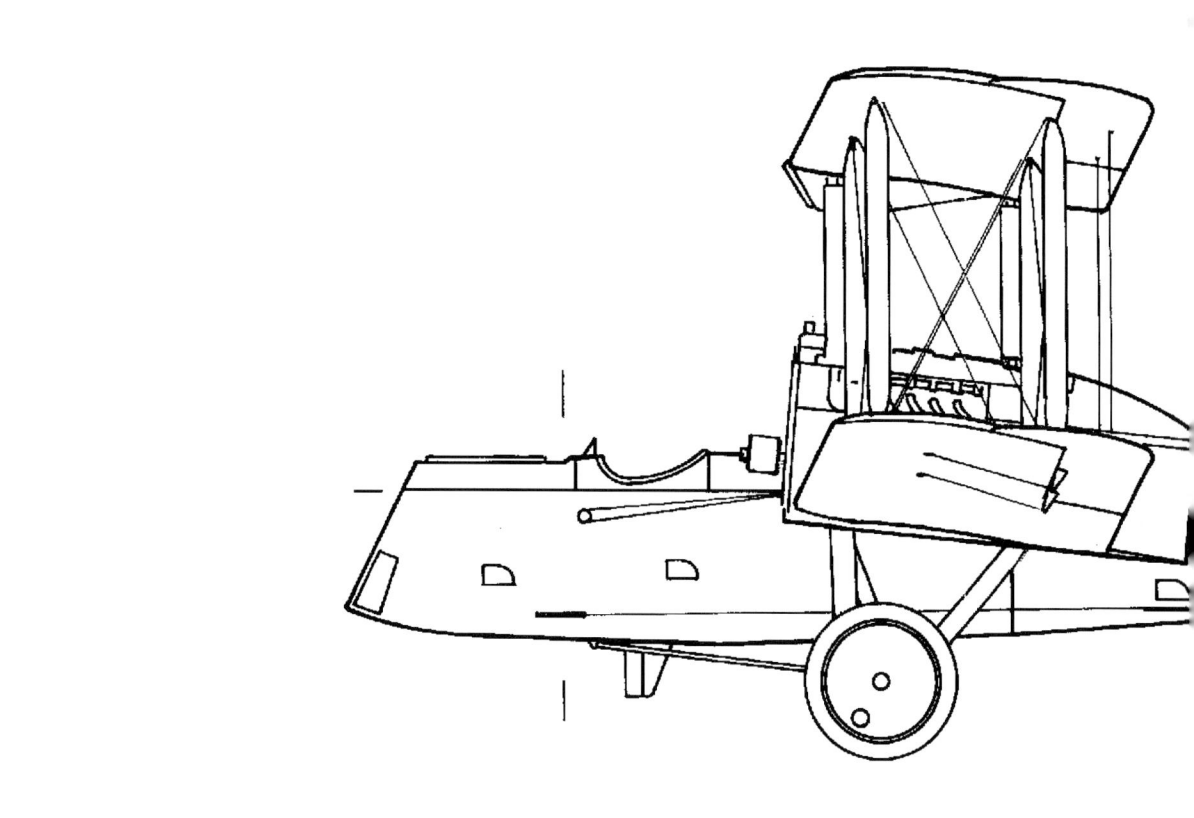

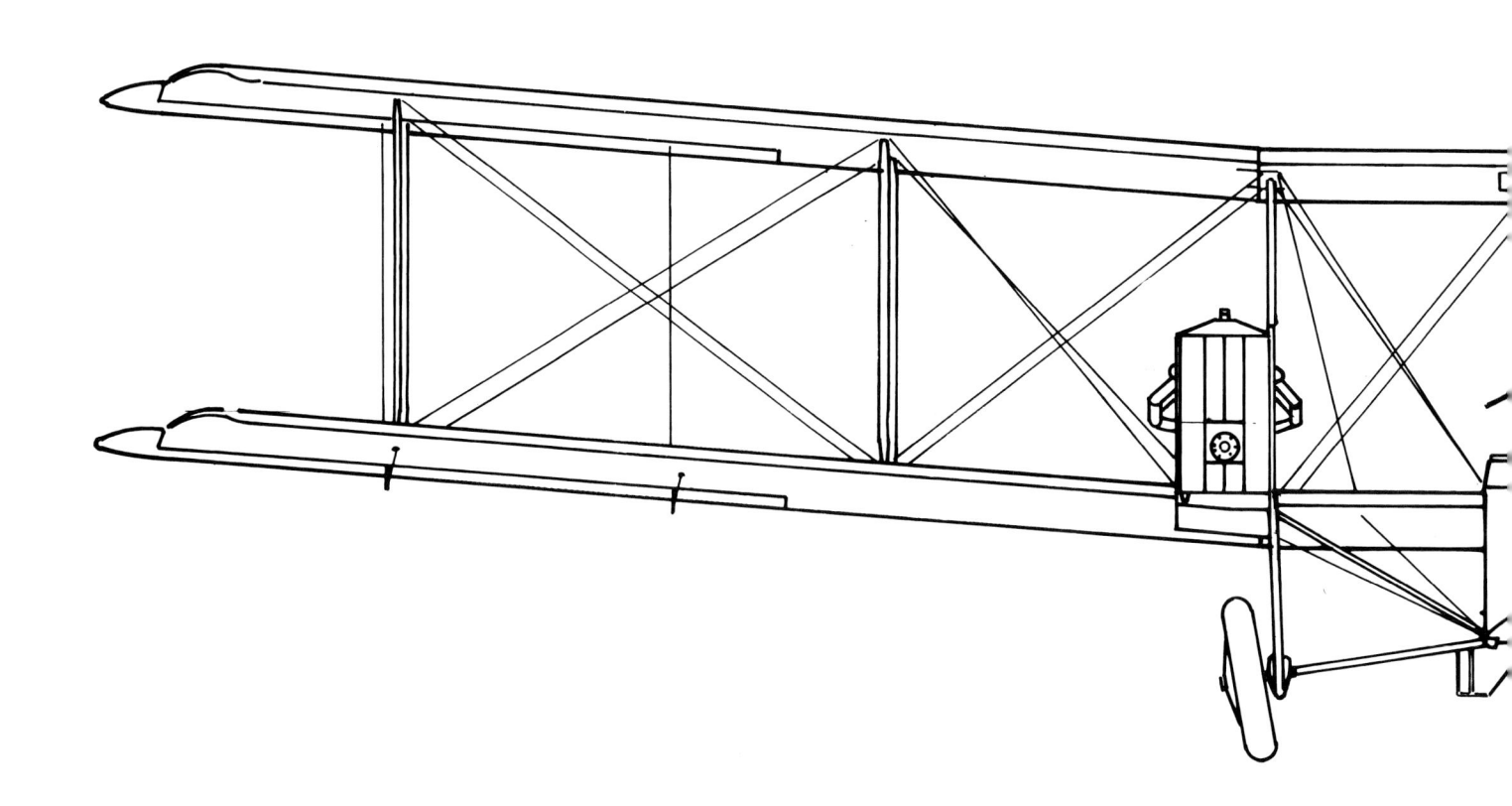

H.10

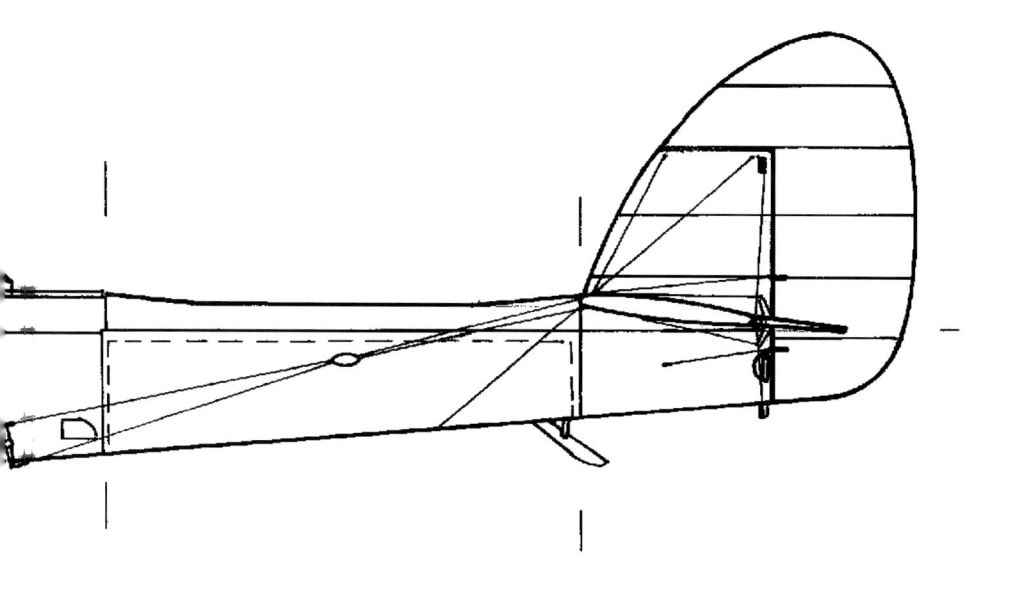

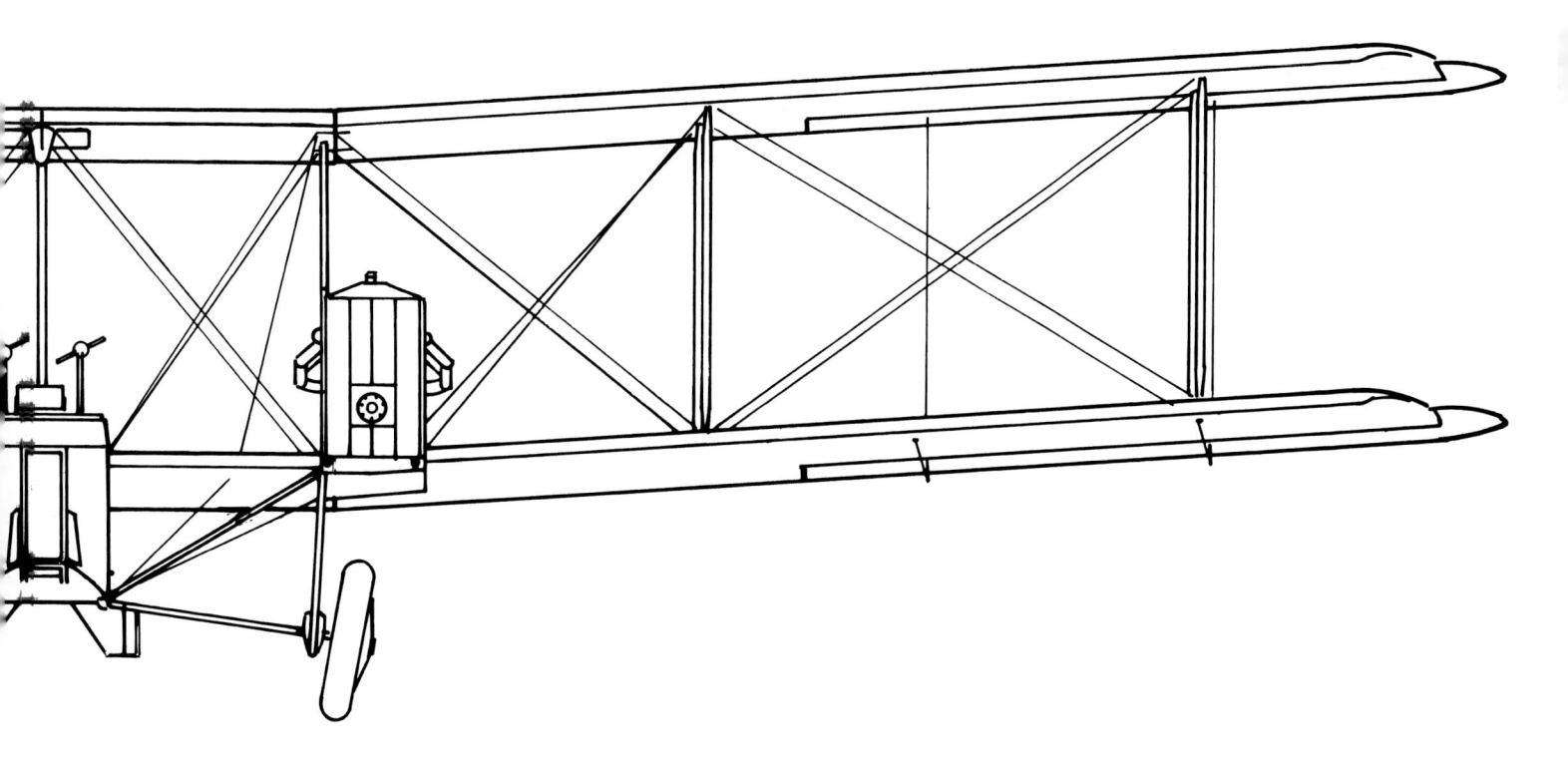

A

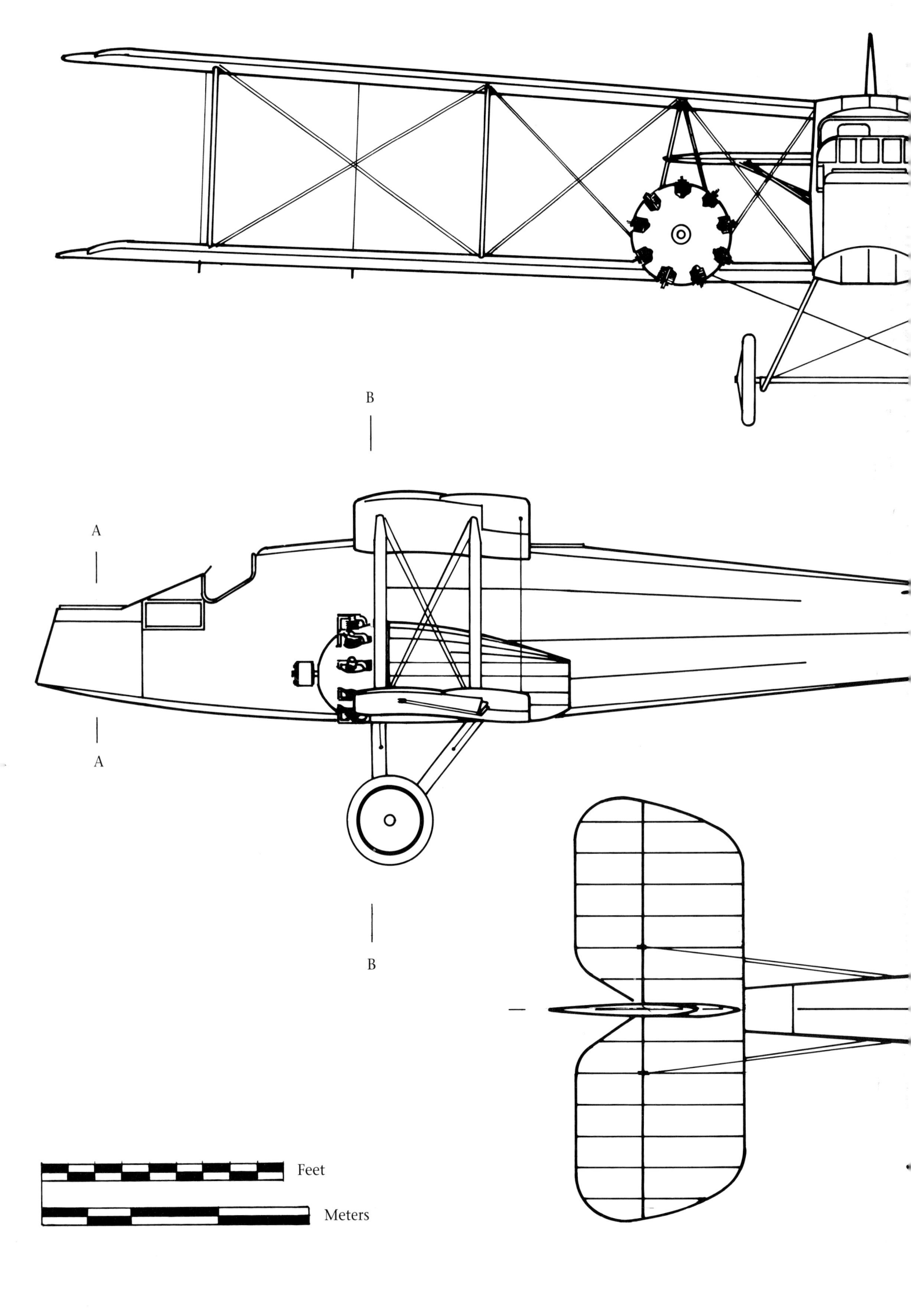

A

B

B

Feet

Meters

Estonia received at 13 D.H.9 bombers (Estonian serials 16, 17, 27 to 32, 67 to 71) in addition to Avro 504K trainers, and Short 184 floatplanes from the British Government in 1919. These were erected by RAF personnel and went into operation with the Aviation Company against the Red Army. In fact the Aviation Company had little influence on the fighting. An Avro 504K flies over Estonian D.H.9 "17" and more 504K trainers while the last aircraft in the line up is a Sopwith 1½ Strutter. Photograph taken at Lasnamägi, Tallin. (Credit: F. Gerdessen)

D9837 carried mixed Estonian and RAF markings when this mishap occurred in the summer of 1918, with RAF cockades to wings and Estonian markings on fuselage and rudder. Known RAF serials for Estonian D.H.9 bombers are D651, D660, D693, D1246, D9837, H9133, H9135, and H9157. These aircraft received the Estonian serials 28, 30, 29, 27, 32, 17, 16, and 31, respectively. Serial 32 survived a long time in Estonian service. (Credit: F. Gerdessen)

Close-up of Serial 32 of the Estonian Air Force on a visit to Finland circa 1926. The aircraft is fitted with skis for operation from snow. (Credit: F. Gerdessen)

Poland was engaged in hostilities with Soviet Russia in 1920 and the D.H.9 saw combat under the Polish red and white checkerboard insignia. The Poles' gallant struggle led to King George V presenting the newly constituted nation with 12 D.H.9 bombers and 20 Sopwith Dolphin fighters. H5721 illustrated here retains its RAF camouflage and serial with the rudder painted red and white and Polish national insignia painted over RAF cockades. (Credit: T. Kowalski)

Estonian ski-equipped Serial 70 in flight. Note the lack of national insignia on the port wing. (Credit: F. Gerdessen)

In late 1921 the De Havilland Aircraft Co. Ltd. modified the D.H.9 to accommodate a third passenger by extending the rear cockpit rearward. This was designated D.H.9C and at least 12 were erected at Stag Lane. Det Danske Luftfartselskab (Danish Air Transport Co) operated four such D.H.9C biplanes, including T-DOBC (ex-H9724), T-DOGH (ex-H9359), T-DODF (ex-H9243). T-DOGH was delivered to Copenhagen on 17 August 1920. It became OY-DIC in 1929. It crashed on 18 August 1930 at Dybvad and was apparently written off. (Credit: RDAF Museum)

H4315 in Polish service in an embarrassing position. Note that the fuselage national insignia is not parallel to the line of flight but the ground angle. (Credit: T. Goworek)

A D.H.9 of the Greek air force circa 1920. (Credit: Hellenic War Museum)

It is often forgotten that real horse power was still very much a commonplace occurrence up to the start of the 1939–45 War. This Danish D.H.9C has the large communal cockpit for the passengers. The windscreen was not very effective and passengers wore helmets and goggles. (Credit: RDAF Museum)

T-DODF with C.C. Larsen as pilot leaves the Clover Field near Copenhagen at the start of the Copenhagen to Hamburg route on 15 September 1920. Note that T-DOGH to the left of the photograph has a communal cockpit for the passengers. (Credit: RDAF Museum)

Further modifications involved the design of a covered cockpit for the passengers and the wings were given 8° of sweep-back. In this form it was still known as the D.H.9C. Two of these later D.H.9C conversions (G-AUED and G-AUEF) were supplied to the Australian company QANTAS for its Charleville to Cloncurry route. QANTAS modified or constructed one at its Longreach, Qld, facility. This D.H.9C, G-AUFM illustrated here, had the pilot's cockpit behind the cabin which must have made it difficult to land and taxi. It served until dismantled in January 1929.

Still preserved in Australia is G-EAQM (ex-F1287) which was flown (or rather crashed) by Ray Parer, with McIntosh as passenger, from the UK to Australia in 1920. The flight was in response to the Australian Government's offer of a prize of £10,000 for the first British aircraft to make the flight. It was won by Ross and Keith Smith and crew in the Vickers Vimy G-EAOU. The D.H.9 left Hounslow, London, on 8 January 1920 and arrived in Darwin, NT, on 2 August 1920 after taking 260 days to complete the 12,900 mile (20,800 km) journey. The initials on the fuselage sides are those of the whisky millionaire Peter Dawson who provided funds for the flight. Despite the crashes, such as that illustrated here when Parer deliberately stalled the aircraft to avoid crowds who ran into the path of the aircraft at Moulmein, Burma. The aircraft was acquired by the Australian Government and after display in the Australian War Museum it was stored, then loaned to Parer's old school where it was displayed in an open sided shed. Due to its deterioration it was regained by the AWM and stored at Duntroon until restoration could commence. When the restoration was almost completed the fuselage and tail were almost completely demolished by a drunk driver who crashed through the wall of the shed in June 1989. The aircraft was again painstakingly restored using as much of the original structure as possible. It may be viewed at the AWM's Trealor Center, Mitchell, ACT, along with aircraft from other conflicts Australia has been involved in.

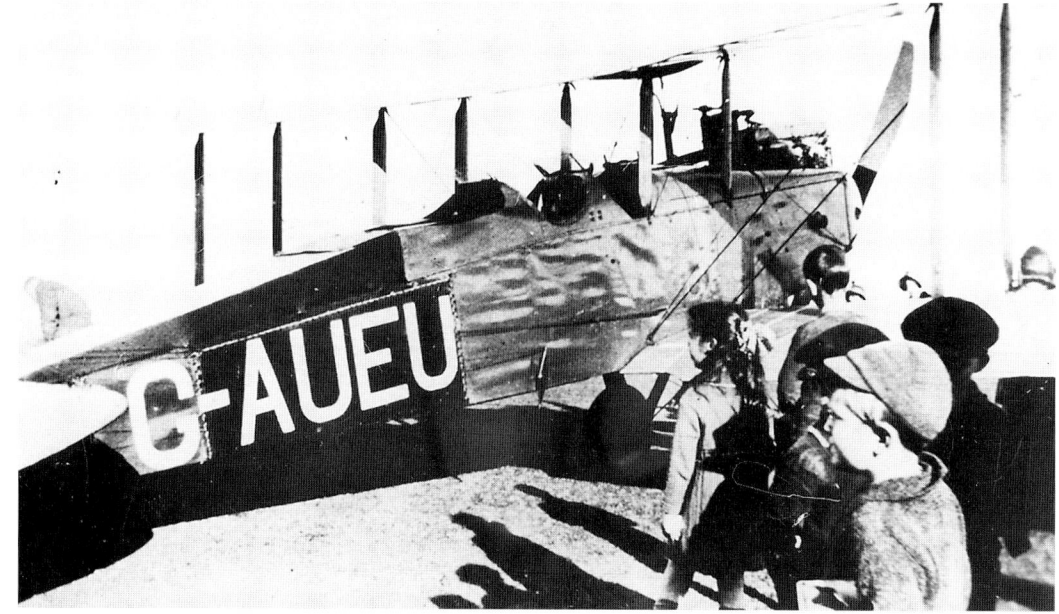

Purchased from the Aircraft Disposals Co. G-AUEU was imported to Australia by Horrie Miller in 1925. Miller joined the RAAF as a Flying Officer in the aircraft repair section at Point Cook and turned up in his own aircraft which was the same as those in the Service, which would be the equivalent to someone owing their own F-111 these days. The machine was offered for sale to the RAAF and the CAS, Richard Williams, recommended that the offer be accepted but the Minister decided otherwise. The aircraft was damaged beyond repair by a gale after landing at port Pirie, SA, on 16 February 1928. (Credit: C. Schaedel)

G-AUHT was Miller's second D.H.9 and received a Certificate of Registration on 28 July 1928. It is suspected that it may have been G-AUEU rebuilt but this has not been confirmed. It is shown with the change in civil registration to VH-UHT after 31 December 1932. Miller's Bristol M1C Monoplane VH-UQI Puck is in the background. The company became McRobertson Miller Aviation Co. Ltd. and disposed of the D.H.9 to Skyways Ltd. of Norwood, SA. (Credit: C. Schaedel)

The Belgium company SNETA made a much more elaborate cabin for two of their D.H.9 biplanes, O-BATA and O-BELG, utilizing the cabin tops and Triplex sliding windows from their D.H.4A biplanes, and fitting underwing luggage containers. O-BELG (ex-F1223) illustrates the hinged rear cockpit hood and access ladder. This aircraft returned to the UK where it became G-EBUN with a CofA issued on 22 December 1927. It was sold to India in January 1929 as VT-AAL. (Credit: via R. Verhegghen)

Below: The end of VH-UHT on 15 May 1937 after its crash at Kadina, SA. Note the racing numerals on the rudder. (Credit: C. Schaedel)

Below: Not many D.H.9 bombers carried squadron markings because these were discontinued as the type entered service. This example bears the white circle of No. 57 Squadron's D.H.4 bombers.

D.H.9A

The failure of the D.H.9 led to a further redesign of the aircraft. As Airco was deeply involved in the design of the D.H.10 bomber, the project was entrusted to the Westland Aircraft Works at Yoevil. Mr. John Johnson was specially loaned by Airco to assist with the project. The new aircraft was to be powered by the 400 hp Liberty 12. In late 1917 large orders were placed in the USA for this engine, as it was impossible to meet the demand for Rolls-Royce Eagle VIII engines.

The new aircraft combined the best features of the D.H.4 and D.H.9 to create the D.H.9A, one of the best strategic bombers of the war. The wing chord was increased, while the tailplane was virtually the same as that of the earlier aircraft. The fuselage was strengthened to take the increased power by incorporating wire cross bracing in the plywood covered sections.

The first D.H.9A was a modified Westland-built D.H.9, B7664. A Rolls-Royce Eagle VIII was fitted to enable trials to take place while awaiting the arrival of the Liberty engines. The armament for the crew was the same as the earlier types; however, double Lewis guns could be fitted for the observer with no loss of performance. Up to 660 lb. of bombs could be carried on external racks under the fuselage and lower mainplanes.

No. 110 was the first squadron to use the aircraft on the Western Front. The type soon proved itself; however, D.H.4 squadrons received the type as replacement aircraft while the D.H.9 squadrons had to struggle on with their inadequate equipment. Nos. 18, 25, 99, 110, 205, 212, and 221 Squadrons were equipped with the type.

The Marine squadrons of the US Navy's Northern Bombing Group, formed in the summer of 1918, utilized the D.H.9A. The Canadian Air Force was just setting up at Upper Heyford as the Armistice was declared. No. 2 Squadron, CAF, (formerly No. 123 [Canadian] Squadron), was equipped with the D.H.9A.

No. 110 was the only Independent Force Squadron to be fully equipped with the D.H.9A. No. 99 had begun to change its D.H.9 bombers with D.H.9A bombers on 4 September 1918, but the squadron did not achieve full replacement before the Armistice. As it was impractical to fly the two types together in formation, only a few operations were flown by the D.H.9A.

The end of the war did not see the end of the D.H.9A's combat career. The British intervention force in Russia included No. 221 Squadron equipped with D.H.4, D.H.9, and D.H.9A bombers with a few Sopwith Camels for escorts. One of their first reconnaissance patrols on 10 February 1919 saw two "Nine Acks" drop their bombs on a parade of cavalry which was displaying red pennants. Unfortunately, it turned out that these were White Russian Cossacks, the side the British were supposed to be helping! The squadron's aircraft were to remain in operation until it was disbanded in September 1919.

No. 47 Squadron was sent to Russia as part of an "advisory force" the British provided to assist the White Russians. Major Raymond Collishaw, DSO, DSC, DFC, was appointed as CO

Two standard D.H.9A biplanes in PC 10 finish with natural metal cowls and radiators.

F982, a D.H.9A of a training squadron in the UK as evidenced by the serial being repeated under the lower wings. F951 to F1100 were ordered on 21 March 1918 under Contract No. 35a/414/C293 from the Westland Aircraft Works. F982 is known to have been with 3 Flying School, Sedgeford, by December 1918.

and its engine as the R.1 and M.5 respectively. The R.1 was used for many years by the Red Air Fleet.

Post-war the D.H.9A formed the backbone of the RAF as a general purpose type in the Middle East. It was in the Middle East where the D.H.9A was to serve a long and distinguished career policing Mesopotamia, Palestine, and Jordan (roughly present day Iran and Iraq). There were always revolts or incursions calling for attention by the RAF. During its long career an additional radiator was fitted and all sorts of miscellaneous gear, from spare wheels to goatskins of water, were attached to the fuselage or between the undercarriage struts.

In addition to its policing duties the RAF initiated future aerial routes. D.H.9A biplanes of Nos. 30 and 47 Squadrons started the Cairo to Baghdad air mail on 21 June 1921. In October 1922 there were eight RAF squadrons in Iraq; four were equipped with the D.H.9A (Nos. 8, 30, 55, and 84).

In 1923 P/Off. Neville Vintcent, with Flt. Lt. J.I.T. Jones, as observer, was one on No.8 Squadron's D.H.9A bombers which bombed and strafed rebel Arabs at Samawah. Unfortunately, Vintcent was forced down by engine failure about two miles (3 km) from the target. A Sopwith Snipe and another D.H.9A saw the plight of the grounded airmen and landed nearby. Flt. Lt. Jack Cottle could only carry one extra man in his D.H.9A so the crew decided to wait until Cottle returned with petrol. As soon as the two British aircraft left the tribesmen started sniping at the airmen from close range. Jones replied with the Lewis gun. When an attack appeared to come from a blind spot, Vintcent would put his shoulders under the 9A's rear and turn the fuselage to give a clear field of fire. It was almost an hour before four aircraft appeared to rescue the crew. Vintcent was awarded the DFC for his actions. The D.H.9A was to be replaced by the Westland Wapiti – an aircraft which used many 9A components!

Canada and Australia received D.H.9A bombers as part of the Imperial Gift. All 11 Canadian aircraft were modified by the Whitehead Aviation Co. and handed over to the Air Board Civil Operations Branch for survey and forestry work. In Australia the type formed the backbone of the RAAF as the D.H.9 biplanes were retired or converted to trainers. The

The size of the D.H.9A's propeller is evident in this view. The thigh boots are worthy of note.

and soon had recruited volunteers to serve with him in south Russia. The squadron was equipped with D.H.9 bombers, but in August began to receive some D.H.9A biplanes, ex-No. 221 Squadron. The war began to go against the White Russians and their British allies. In February 1920, Collishaw's D.H.9A was hit by ground fire and he lost his engine. Forced to land, he managed to coax the damaged engine to taxi some 20 miles (32 km) to safety. The surviving aircraft were handed over to the White Russians when the British withdrew in late March, early April. It continued in combat with the RAF after the Armistice in Russia. The Soviets recovered some examples and copied it

F1072, aircraft "B", another standard Westland-built D.H.9A. It is known to have been with No. 18 Squadron by November 1918.

Comparison of the D.H.9 and D.H.9A

The D.H.9 and 9A were compared by the official British historian in operations held over 28 to 30 October 1918. The D.H.9A bombers of No. 205 Squadron were given no fighter escort and assigned the railhead at Namur as their objective. The three D.H.9 squadrons (Nos. 27, 49, and 107) were allotted one two-seat and three single-seat fighter squadrons for protection both ways to the limit of the fighter's range.

No. 205 reached its objective each day at its operational altitude of 17,000 feet (5,180 m) and had little in the way of attacks from enemy aircraft. The D.H.9 bombers on the other hand flew at about 13,000 feet (3,960 m), which was closer to their effective ceiling than official performance figures of 18,000 feet (5,490 m), and were subject to continual fighter attacks.

When attacked the D.H.9 biplanes dropped their bombs with little being achieved. The formations were depleted by engine trouble causing aircraft to drop out of the formations. On the 30th Namur was again bombed by the D.H.9A bombers of No. 205 Squadron. No. 107 Squadron's D.H.9 bombers, escorted by S.E.5a fighters of No. 1 Squadron and Sopwith Snipes of No. 43 Squadron, alone of the D.H.9 squadrons reached and bombed its objective at Mariembourg. In the three days the D.H.9A bombers bombed the most distant objective on all days without loss. Even with fighter escort only one squadron of D.H.9 bombers reached its objective with the following losses by unit:

On 28th
    No. 27   Lost D535 with the crew POWs.
                  D1092 was damaged but the crew was OK.
    No. 49   D3223 was damaged but the crew was OK.
On 29th
    No. 107   One D.H.9 was lost with the crew killed.
On 30th
    No. 49   D502 was lost with the crew killed.
                  D3260 was shot up but the crew returned OK.
                  D3265 was shot up but the crew returned OK.
    No. 107   One D.H.9 crew was wounded.

D.H.9A was the longest serving Imperial Gift aircraft in the RAAF, the last being approved for write-off in February 1930.

The D.H.9A did not have an extensive civil career unlike its predecessors. A number were used for civil work such as the Lion-powered D.H.9A mailplanes of Aircraft Transport and Travel Ltd. on the Cologne mail service in 1919–20. The D.H.9R, G-EAHT, was a sesquiplane racer erected from D.H.9A components. It broke several British speed records at Hendon in November 1919.

The stocks of D.H.9A parts were so vast that the type was kept in service. Westland produced 36 Walrus fleet spotter aircraft from D.H.9A components. The Westland Wapiti, the successor to the D.H.9A, originally used D.H.9A wings and tail surfaces.

G-EBAN was fitted with a Rolls-Royce Eagle VIII and demonstrated in Spain in February 1922. Its success at the Military Trials led to a small batch of such aircraft being used by the Spanish Air Force.

E9703 in factory fresh markings. Part of a batch of 100 (E9657 to E9756) built by Mann Egerton, this aircraft displays the gray applied to metal and ply panels with only fabric surfaces being finished in PC 10. Lower surfaces of all wings and tailplane were clear doped. This aircraft served in the RAF post-war, being recorded at the Parking Depot Ascot and at the Air Depot Aboukir as well as 4 Flying Training School. (Credit: via P.H.T. Green)

It took a lot of man power to pull over the 400 hp Liberty engine of the D.H.9A as evidenced in this view. (Credit: Late F.S. Briggs)

Above and below: Close-up of the D.H.9A cockpit.

Right: D.H.9A "B" of No. 110 Squadron, RAF.

Below: D.H.9A of No. 110 Squadron, RAF.

Below: D.H.9A "L" of No. 205 Squadron, RAF, rests in the snow.

Line-up of D.H.9A aircraft of No. 205 Squadron, RAF.

D.H.9A E8538 of "C" Squadron, USMC, assigned to the Northern Bombing Group. These aircraft were powered by Packard-built or Lincoln-built Liberty engines. (Credit: National Archives)

USMC D.H.9A and D.H.4 aircraft at La Fresne aerodrome, fall 1918. (Credit: U.S. Marine Corps Historical Center)

Below: Aircraft "G" of No. 205 Squadron, RAF, was E1008. Lt. K.G. Nairn is in the pilot's cockpit while Lt. K.J. Golding is standing in the Scarff ring of the rear cockpit. Note the snow speckled wings. Squadron markings had been discontinued by this time and the individual aircraft letters were styled to reflect some squadron identity. (Credit: David McGuiness collection)

Above: There is only one surviving D.H.9A remaining today. F1010 was captured on 5 October 1918, when the No. 110 Squadron aircraft was brought down together with three others by ground fire and aerial attacks. 110 Squadron had been equipped by His Serene Highness, The Nizam of Hyderabad, and F1010 bore an appropriate inscription as indeed did most of the Squadron's aircraft.

Exhibited at the *Deutsche Luftfahrt Sammlung* in Berlin, it was transferred from Berlin following Allied raids which severely damaged the museum in November 1943. Transferred to Czarnikau (today Czarnkow, Poland) for safety, the surviving aircraft were brought to the Polish Muzeum *Lotictwa i Astronautyki* in Krakow but most were not on public display. The remains of F1010 were eventually exchanged with the RAF Museum for a Spitfire XVI in 1977. The Museum has since built a new set of wings for the aircraft and restored it to as new condition. It is depicted in the condition it was received in from Poland. (Credit: M. Krkyzan)

Below: H3510 in post-war service. Aircraft is finished in V.84 aluminum overall. Note the extra radiator, enlarged gravity fuel tank, supply carrier and spare wheel under the fuselage, and Holts flare holder under the lower wing. The tailplane appears to be painted red as is the individual aircraft letter "L" when in No. 8 Squadron service.

This aircraft was part of a batch of 150 (H3396 to H3495) from Westland Aircraft Works. All these aircraft were delivered between November 1918 and August 1919. H3510 was recorded at the Air Depot Hennaed; was with No. 8 Squadron from February 1925 to 1926; at No. 84 Squadron during November to December 1927, and then with No. 55 Squadron. (Credit: H.G. Crone via J.M. Bruce/G.S. Leslie collection)

J6957 was a rebuilt D.H.9A by Westland to Contract No. 375546/2 of 14 March 1923. Fitted with a 465 hp Napier Lion II engine it was at the ACE on 24 April 1923, and then to the RACE by 7 August that year with a metal airscrew as shown here. Note the oleo undercarriage and bomb. It was re-engined with a 400 hp Liberty some time after mid-1925, and was back at the RACE on 23 June 1926 with steel wings and oleo undercarriage. On 30 August 1927, the aircraft was involved in a tragedy when the passenger's parachute became entangled with the aircraft. Last flown 6 December 1927. (Credit: J.M. Bruce/G.S. Leslie Collection)

Right: D.H.9A E8754 tropical with extra radiator, full supply carriers under wings, spare wheel under fuselage, and extra gravity tank on top wing. The aircraft is stated to be in No. 14 Squadron service although this has not been confirmed from documents.

E8754 was one of 400 D.H.9A bombers (E8407 to E8806) ordered in March 1918 from the Aircraft Manufacturing Co. Ltd. All were delivered by March 1919. The aircraft went to the Air Depot at Hennaed and was recorded as aircraft "O" at No. 8 Squadron in Augusts 1923. It was the AOC Iraq's aircraft in 1924. It was returned to the Air Depot then to the UK where De Havilland reconditioned the airframe on Contract No. 579165/25. At No. 60 Squadron by March 1927. Suffered a forced landing due to engine failure on 4 December 1928. Repaired, it was back with No. 60 Squadron by December and served till February 1929. From here its service becomes confused. It is recorded as at No. 27 Squadron from July to October 1929, and also as aircraft "Y" of No. 60 Squadron from June 1929 to February 1930. It is last recorded as with No. 27 Squadron from March to May 1930 when it is thought it may have been fitted with dual control. (Credit: H.G. Crone via J.M. Bruce/G.S. Leslie Collection)

E961 in service in the Middle East circa 1926. Note the additional radiator, spare wheel under fuselage, and bombs on the wing carriers. The wing tips are painted red for easy visibility in the event of a forced landing. Built by the Whitehead Aircraft Co. as part of batch E701 to E1100, only 40 were delivered, the rest being cancelled in November 1918. This aircraft was at the Air Depot Hennaed; at No. 8 Squadron by September 1923, but back at the Air Depot the next month. In June 1924 it is recorded at No. 30 Squadron; then at the Air Depot Aboukir and at No. 4 Flying Training School. (Credit: P.H.T. Green)

Canadian D.H.9A on skis. All 11 Imperial Gift Canadian D.H.9A biplanes were operated in civil guise by the Canadian Air Board's Civil Operations Branch. Note the rear cockpit conversion to dual control. (Credit: via J.M. Bruce.)

Australia was given 30 D.H.9A bombers as part of the Imperial Gift. One, E8616, was lost on 23 September 1920, with Capt. W.J. Stutt and Sgt. A.G. Dalzell while carrying out a search for a missing schooner, the *Amelia J*. Another D.H.9A was purchased so that 30 D.H.9A bombers received RAAF "A" serial numbers.

Depicted here are A1-17 and A1-26, typical of the reconditioned aircraft operated by the RAAF. They are painted in V.84 aluminum overall with black serials and a thin white outline to the roundels.

A1-26 was ex-RAF E8590. Received by the Australian Air Corps (AAC) in November 1920 it was stored until June 1922, going to "E" Flight at 1 Flying Training School (FTS) the following month. The machine suffered a casualty in September 1923 but was back in service by at least October the following year. A1-26 took part in the disastrous RAAF Pageant practice at Flemington Racecourse in 1924. Flt. Lt. A. Hempel made a good approach and landing but not having the benefit of brakes the D.H.9A ran towards the boundary fence until Hempel saw he had no hope of stopping and tried to ground loop. In the ensuing cloud of dust the biplane came to rest minus its undercarriage and sundry pieces of wood and fabric. The crew was unhurt physically but suffering from the comments of abattoir workers who were watching the proceedings from the meatworks opposite the racecourse.

F/Off. McCauley with F/Off. Duncan suffered a casualty around April 1925, a Board of Survey being held at 1 Air Depot that month. The aircraft was back in service the next month and being reconditioned at 1 FTS the following February. It was in service until May 1929 after hitting a light standard at Laverton. The Air Board approved disposal of the airframe and Liberty 65074 the same month. On 26 June 1929, 1 Air Depot confirmed conversion completed.

A1-17 was ex-F2779. Received in May 1920 it was one of the few D.H.9A biplanes operated by the AAC, being tested on 24 May by Capt. Adrian T. Cole. The aircraft performed miscellaneous duties during this period, as the AAC had no direction. On 27 May, Cole made flights over various Melbourne suburbs to show General Blamey, the Army Chief of Staff, around. The following month on the 17th, Cole and Capt. De La Rue made a height test, reaching 27,000 ft before being forced down by lack of oxygen. This was an Australian altitude record. On 17 October 1920, it was transferred from "A" Flt. to the Aircraft Repair Section with Liberty WD65071/MN25769/2084. On formation of the RAAF it was taken on charge 1 FTS from Central Flying School, AAF. From 1 July 1922 to 1923 it was on the strength of 1 FTS. On 7 February 1924, A1-17 suffered a casualty while being flown by F/Off G.H. Kindrer, being subject to a Board of Survey on 6 March at 1 Air Depot. From April to August 1924, it was on strength of No. 1 Station. The following year it was recorded on strength of 1 Flight Training School. In July an in flight breakage of the rudder was documented. A1-17 is next recorded in August 1926 when it was being used for instruction, presumably having being converted to dual. On 16th Cdt. Frewin suffered a crash, the aircraft going to the Air Depot for storage. By July the following year it was being reconditioned at Pratt Bros. This took an inordinate amount of time. It was back in service with No. 1 Squadron by at least February 1929. On 3 May 1929, the machine suffered damage to the airscrew, undercarriage and radiator when F/Off. Brelaz and P/Off. F.W. Thomas allowed the aircraft to turn out of wind when landing at Laverton. The following month saw a forced landing at Keillor, Vic when P/Off. Fleming with AC1 Stevens operated the radiator incorrectly. The aircraft was on a flight to Sydney with three others. The 7th July saw F/Off. Stevens suffer a forced landing at Point Cook when the big end bearing collapsed. The aircraft was not long back in the air when it suffered another forced landing on 31 October, this time at the hands of P/Off. G.W. Boucher when the camshaft drive broke, causing engine failure. At some time the airframe was given a second reconditioning by Matthews and Hassell and was fitted with the late slotted wing. Ministerial approval was granted on 4 February 1930, for the writing off charge and destruction of all remaining D.H.9A biplanes, together with all spare parts, and 13 Liberty engines to be offered for sale, thus bringing an end to the last of the Imperial Gift machines in the RAAF.

This photograph shows four of the RAAF's D.H.9A biplanes at Wave Hill with Les Holden's D.H.61 *Canberra*. The D.H.9A biplanes have been given oleo landing gear and slotted wings in an attempt to keep them reasonably up to date.

In 1929 when Charles Kingsford Smith and his crew went missing in the *Southern Cross*, the RAAF procrastinated and did not participate in the search. One search was mounted by Smithy's friends Keith Anderson and Bobby Hitchcock in Westland Widgeon VH-UKA named *Kookaburra*. On 12 April 1929, Les Holden in his D.H.61 Canberra found the missing *Southern Cross*. In the jubilation over finding Smithy safe and sound, the fact that the *Kookaburra* was missing was overlooked.

A public outcry was soon raised over the missing *Kookaburra* and the federal government, which had been tardy in committing itself to search for the *Southern Cross*, promptly made RAAF aircraft and crews available to search for it. On that same day that Holden found the *Southern Cross*, D.H.9A biplanes A1-1 and A1-7 of 1 Squadron under the command of Flt. Lt. Charles Moth Eaton, had left Laverton for Alice Springs to search for the lost aircraft. The D.H.9A biplanes were sent as they were the only RAAF aircraft available with sufficient endurance to operate in the Outback.

A second contingent comprising F/Off Ryan with Corp. E.N. Dean in A1-5, Sgt. Douglas and AC1 Smith in A1-20, and F/Off. Douglas and AC1 Allen in A1-28, left on the 17th. One of these aircraft was equipped with a shortwave wireless transmitter, and the other two were fitted with standard RAAF apparatus for dropping supplies of water and food by parachute.

Eaton's party reached Alice Springs on the 16th at 6:45 pm. The three additional D.H.9A bombers under Ryan arrived in Oodnadatta on the 18th. The area of the Northern Territory where the *Kookaburra* went missing was almost devoid of physical features that are of any assistance to pilots in the matter of locating their position during the course of patrols.

The age of the de Havillands was also a problem. One account of the affair records that it was standard practice as part of the pre-flight inspection to check the amount of white lead in the oil, and from this deduce whether anything serious was wrong with the engine. This account also gives the absolute life of the Liberty engine at this time as 400 hours.

On 21 April Eaton crashed. The 22nd saw approval to burn the wreck where it lay was telegraphed to Eaton and this was done after all instruments, etc., had been salvaged.

Eaton had been coming up to Wave Hill with his RAAF contingent, systematically searching on the way. QANTAS was approached to help and their D.H.50J G-AUHE *Atalanta* flown by Capt. L.J. Brain was dispatched from Brisbane in Queensland to help in the search. Brain was to meet the RAAF contingent at Wave Hill. He had left Newcastle Waters on the 21st when he saw smoke. Knowing that the Tanami Desert was completely devoid of life due to the total absence of water, Brain flew to investigate and discovered the *Kookaburra*. One body could be seen under the wings and a gallon container of water was dropped before the D.H.50J proceeded to Wave Hill.

Eaton had sent Ryan ahead to Tennants Creek and Newcastle Waters. On the way Ryan had seen the smoke and communicated with Eaton. On questioning Ryan on the amount of white lead in his oil and receiving a reply that it was an excess amount, Eaton had refused permission for Ryan to investigate the smoke until the rest of the contingent arrived, because if he got into trouble there would be two aircraft lost. This was the correct decision but it robbed the RAAF of the honor of finding the *Kookaburra*.

The *Atalanta* and Air Force contingent rendezvoused at Wave Hill and it was considered that "the risk in flying 9A's, which boiled when they came down low, over such country, was not justified," and so Eaton led a ground party to the site of the *Kookaburra* using horses and a lorry. Anderson was found some 400 meters from the plane where he had perished in an attempt to walk to water. Both bodies were buried where they lay and the party trekked back.

The RAAF's troubles were not over. A1-28 was unserviceable at Wave Hill and on 7 May, Eaton telegraphed that A1-20 had suffered an engine fire. Both aircraft were struck off charge on the 9th after salvaging "prop, radiator, plugs, generator, and distributor." A1-5 and A1-7 were reported at Maree on their way to Adelaide on the 10th. The RAAF had lost three aircraft and the rescue attempt was the subject of criticism in the press. Brain thought otherwise and wrote to the Secretary of the Air Board that *knowing the old machines with which they are equipped, the class and amount of country over which they have flown, the severe flying conditions, and the spirit displayed by them in their work, I should like to commend the Air Force party engaged on the search to the greatest praise and special attention.*

The government did not feel that Eaton and his crews deserved special attention and no awards were made to the RAAF contingent at this time.

Above and above right: D.H.9A taking off; an individual letter "A" is on the nose.

Above and below: F1021 was a Westland built D.H.9A (Batch F951–F1100). As aircraft "W" of No. 101 Squadron, Independent Force, it was one of 13 D.H.9A bombers which set off for Frankfurt. Clouds split up the formation and none bombed their objective. Only five aircraft managed to bomb the alternative target, Köln. The D.H.9A biplanes were set upon by masses of German defenders and seven were lost. F1021 was fortunate in that 2/Lt. P. King and 2/Lt. R.G. Vernon were taken POWs, although Vernon later died of wounds. Their aircraft is shown in German hands. Note the lack of rudder and twin Lewis guns. (University of Texas)

The first aircraft in this line up (?439) appears to have the squadron badge of No. 110 Squadron on its nose. There is a numeral (10?) marked in white on the fuselage behind the cockade. H3485 is second in line and no badge is visible. The polished metal cowlings show up in this photograph. (University of Texas)

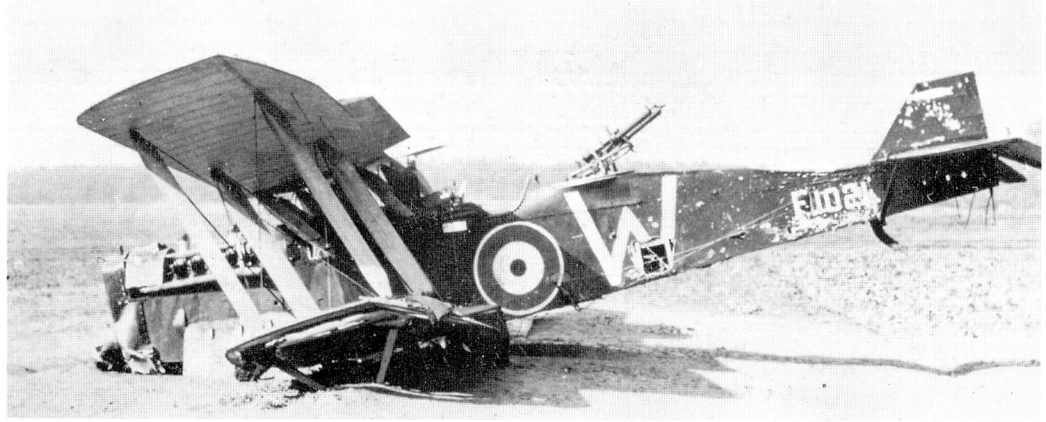

Above: This RAF D.H.9A in unfortunate circumstances shows the auxiliary radiator, badly bent in this case, and spare wheel attached to the fuselage bottom. The D.H.9A was the workhorse of the RAF in India and the Middle East.

The USD-9A

Initially designated USD-9, the USD-9A was an American redesign of the D.H.9A to replace the Liberty D.H.4. It was to be used as a day and night bomber as well as for reconnaissance. The prototypes were built by the Engineering division workshops with orders being placed for at least 4,000 aircraft with US firms.

The Special British Aviation mission to the USA found that the USD-9A was "similar to the English D.H.9A but possesses certain well thought-out alterations in detail" including a supplementary retractable radiator. Performance was thought to be the same as the British machine if the weight was kept close to that of the original.

Flight testing commenced in July 1918 at McCook Field. A larger more rounded rudder was fitted at one time. Only about five were completed, one, A.S.40118, being a single-seater with a pressure cabin.

The end of the war led to cancellation of the project and it is not known if any ever reached the European Theatre, although two are recorded as being crated for shipping overseas.

Dayton-Wright built USD-9A serial 40042 with the McCook Field number P-40 on the rudder. This aircraft survived until 30 November 1920 when it was written off after a fire.

USD-9A front view shows the retractable auxiliary radiator which was located just forward of the front undercarriage strut. This feature was not seen on wartime British D.H.9A bombers. This aircraft became McCook Field P-64 and was in service until early 1922.

Above, above right, and right: Views of unmarked USD-9A.

USD-9A in national markings. Note the retractable auxiliary radiator. McCook Field photo 1595.

USD-9A serial 40044 in full national markings. (Credit: USAF Museum)

D.H.10 Amiens

The D.H.10 was a larger version of the D.H.3. Of similar construction to its predecessor, the airframe structure was built of spruce and ash with fabric and ply covering. The front portion of the fuselage was a plywood covered box-like structure with a rear portion comprising the usual warren girder construction. Rudder and elevators had steel tube trailing edges. The engine support struts were also of steel tube faired by wooden formers covered in fabric. Two 230 hp B.H.P. water-cooled engines were mounted as pushers. Provision was made for a front gunner, pilot, and rear gunner behind the wings. The rear gunner's cockpit had full dual controls.

The prototype made its first flight on 4 March 1918. Performance was down on predictions and so the next aircraft had engines of greater power installed as tractors, 360 hp Rolls-Royce Eagle VIII engines being installed in C8659 while C8660 had 400 hp Liberty 12 engines. This latter was to be the production version. The Air Board had given the D.H.10 the name Amiens, so that the first two were the Amiens Mk.I and Amiens Mk.II respectively with the production aircraft being the Amiens Mk.III.

Although orders were placed for substantial numbers of D.H.10 bombers, only eight were on RAF charge when hostilities terminated, two or three of which had reached No. 104 Squadron at Maisoncelle, France. The Mann, Egerton and Co.-built D.H.10A Amiens Mk.IIIA saw the Liberty engines mounted directly on the lower wing.

The type, along with the Vickers Vimy, went on to serve in the post-war RAF. It was used by Nos. 60, 97, 104, and 216 Squadrons in India, France, and Egypt respectively. No. 60 Squadron's D.H.10 bombers were involved in operations

This view of a D.H.10 prototype (probably C8658) shows the "Germanic" installation of the pusher engine nacelles. Note the rectangular hole in the rudder balance. This unusual feature appeared on production D.H.10 bombers. (Credit David McGuiness collection)

The serial F1?? indicates that this is one of 16 Amiens Mk.III aircraft (F1867–F1882) ordered on 27 March 1918 under Contract No. 35a/509/C385 from the Aircraft Manufacturing Co. All were delivered by March 1919. The original Airco caption identifies this as a D.H.10A with two 400 hp Liberty engines mounted directly on the bottom wing. This was F1869, the prototype of the D.H.10A. (Via J.M. Bruce)

against Pathan tribesmen in April 1920. In November they were again involved in attacks against rebels together with Bristol Fighters and D.H.9A bombers. By 1922 the RAF in India was in a poor condition due to shortage of spares due to chronic underfunding. The austerity measures of the government saw the RAF as well as the other defense services starved of funds. In February 1923 No. 60 had only two unserviceable D.H.10 bombers still on strength and the type was withdrawn by April.

In Egypt the type served for two years, replacing the Handley Page 0/400 bombers in No. 216 Squadron. They continued the Cairo to Baghdad air mail which had been started by D.H.9A biplanes. The radiators were increased in area for the hot climate and some of No. 216 Squadron's D.H.10 biplanes had an extra cockpit just behind the pilot's cockpit. The D.H.10 could make the round trip from Cairo to Baghdad in a single day; 10 hours from Cairo and about 9 hours return. The route was eventually taken over by Imperial Airways and their De Havilland Hercules airliners in 1926.

Although never used in its intended task, the D.H.10 served faithfully under conditions which were never contemplated when it was designed. It deserves its place in RAF history.

E9057 was part of a batch of 150 D.H.10 Amiens Mk.III bombers ordered from the Daimler Co. Ltd. and constructed at the National Aircraft Factory No. 2 under Contract 35a/432/C313 dated 20 March 1918. E9057–E9096 were delivered between 11 January and 21 June 1919, the rest being cancelled in November 1918. This view shows the installation of the 400 hp Liberty engines above the lower wing. The front part of the fuselage is doped gray while the rest of the airframe is PC 10 with clear doped undersurfaces. (Credit: J.M. Bruce/ G.S. Leslie Collection)

E6040 of No. 60 Squadron, being bombed up for a raid on Dhatta Khel, India, 12 January 1922. Note bomb carrier under wing. This aircraft is reported as having been crashed on 6 January which throws some doubt on the date stated in the caption. It was with No. 90 Squadron in 1920 when that unit was renumbered as No. 60. (Via J.M. Bruce)

E5450 in dark camouflage (PC10/PC12). This Amiens Mk.III was with No. 97 Squadron in January 1919. It is also reported as having crashed on 6 January 1922 when landing at Risalpur, India.

D.H.10A E5557. 200 D.H.10 Amiens Mk.III aircraft were ordered under Contract No. 351/427/C314 on 20 March 1918 from the Aircraft Manufacturing Co. E5437–E5558 were delivered, with the rest being cancelled in June 1919. The possibility of a threat to the supply of Liberty engines led to the substitution of two 375 hp Rolls-Royce Eagles with the designation changed to D.H.10C. E5557 and F8441 were definitely D.H.10C biplanes. E5557 saw service post-Armistice with Aircraft Transport and Travel in company with D.H.10 E5488 (registered as G-EAJO). (Via J.M. Bruce)

E7851 was a Amiens Mk.III built by the Siddeley-Deasy Motor Car Co. Ltd. It was serving with No. 216 Squadron in 1923 as shown here. It is finished V.84 aluminum overall, note the uncowled engines. It had another cockpit installed behind the normal one. E5507 of the same squadron also had the second cockpit. It has been suggested that it was for a relief crew during the Cairo to Baghdad flights. (Via J.M. Bruce)

Tropical radiators are well displayed by this D.H.10 of No. 216 Squadron in Egypt. Note the individual marking of a dark (red) diamond on the white nose stripe. (Via J.M. Bruce)

D.H.11 Oxford

Three D.H.11 Oxford twin-engined bombers were ordered on Contract No. 351/2150/C2486 of 24 July 1918. Construction of H5892 and H5893 was discontinued by 30 June 1919 and these serials were reallocated to the Sopwith Buffalo. The solo example produced, H5891, did not fly until January 1920.

The D.H.11 was a neat three-seat biplane with the top wing mounted flush with the upper fuselage. Intended for long range fighter reconnaissance missions or short and long range daylight bombing, the Oxford was an elegant aeroplane. Construction was similar to the D.H.10 with a fabric-covered wooden framework with steel tubing used for highly stressed areas such as engine mounts. Power was supplied by two 320 hp A.B.C. Dragonfly engines mounted directly onto the lower mainplanes.

The prototype was half completed when the Armistice slowed construction work. By mid November it was decided that the problems with the Dragonfly engine were not going to be overcome in the short term and high-compression Siddeley pumas would be fitted whereupon the aircraft would be known as the Oxford Mk.II. In the event the aircraft emerged with Dragonfly engines. It apparently did little flying, undergoing manufacturers trials in October to November 1920. It soon disappeared, the Dragonfly engine causing it to lapse into obscurity.

Left and below: The sole example of this elegant bomber built. (Credit J.M. Bruce/G.S. Leslie)

D.H.14

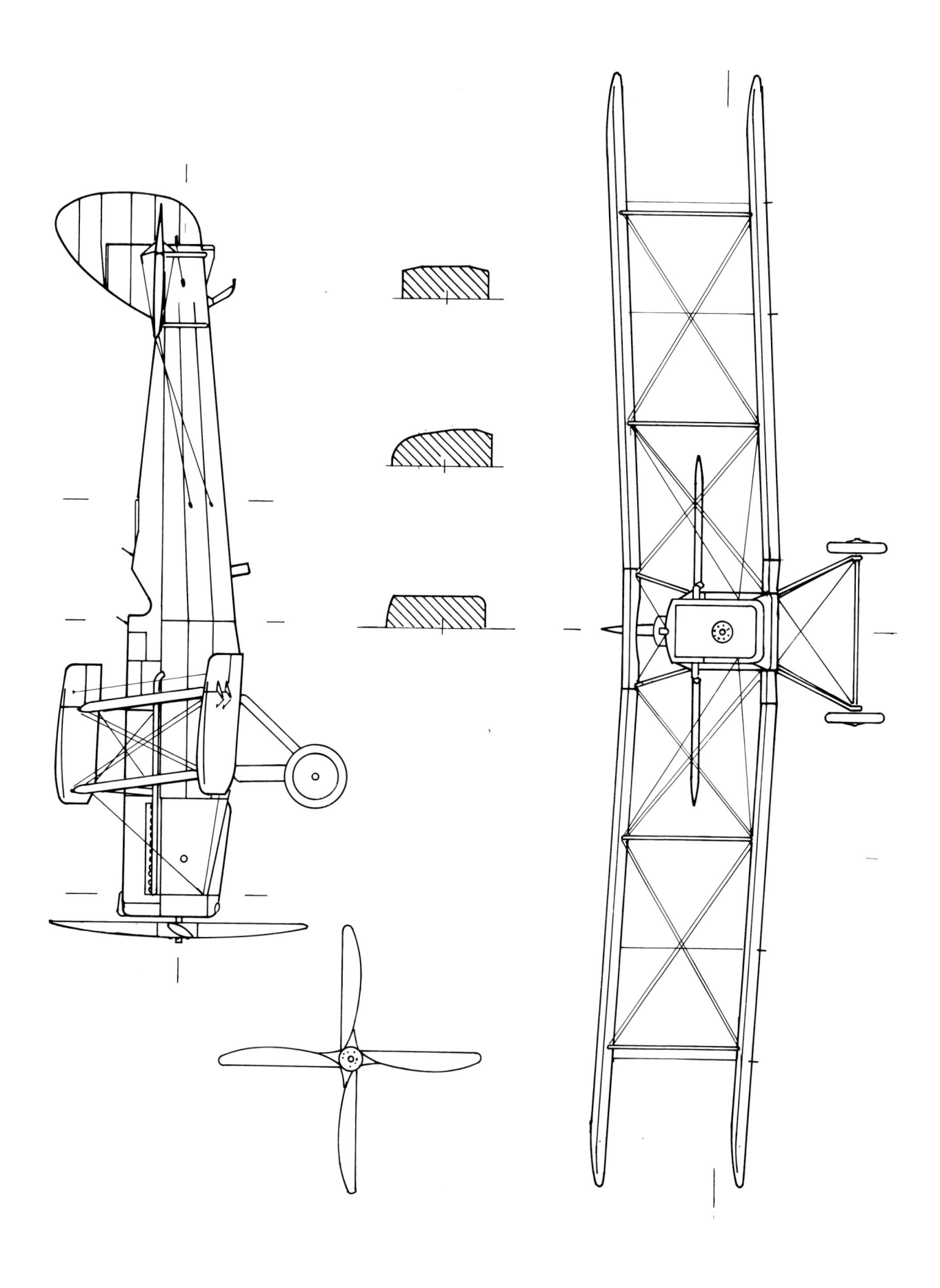

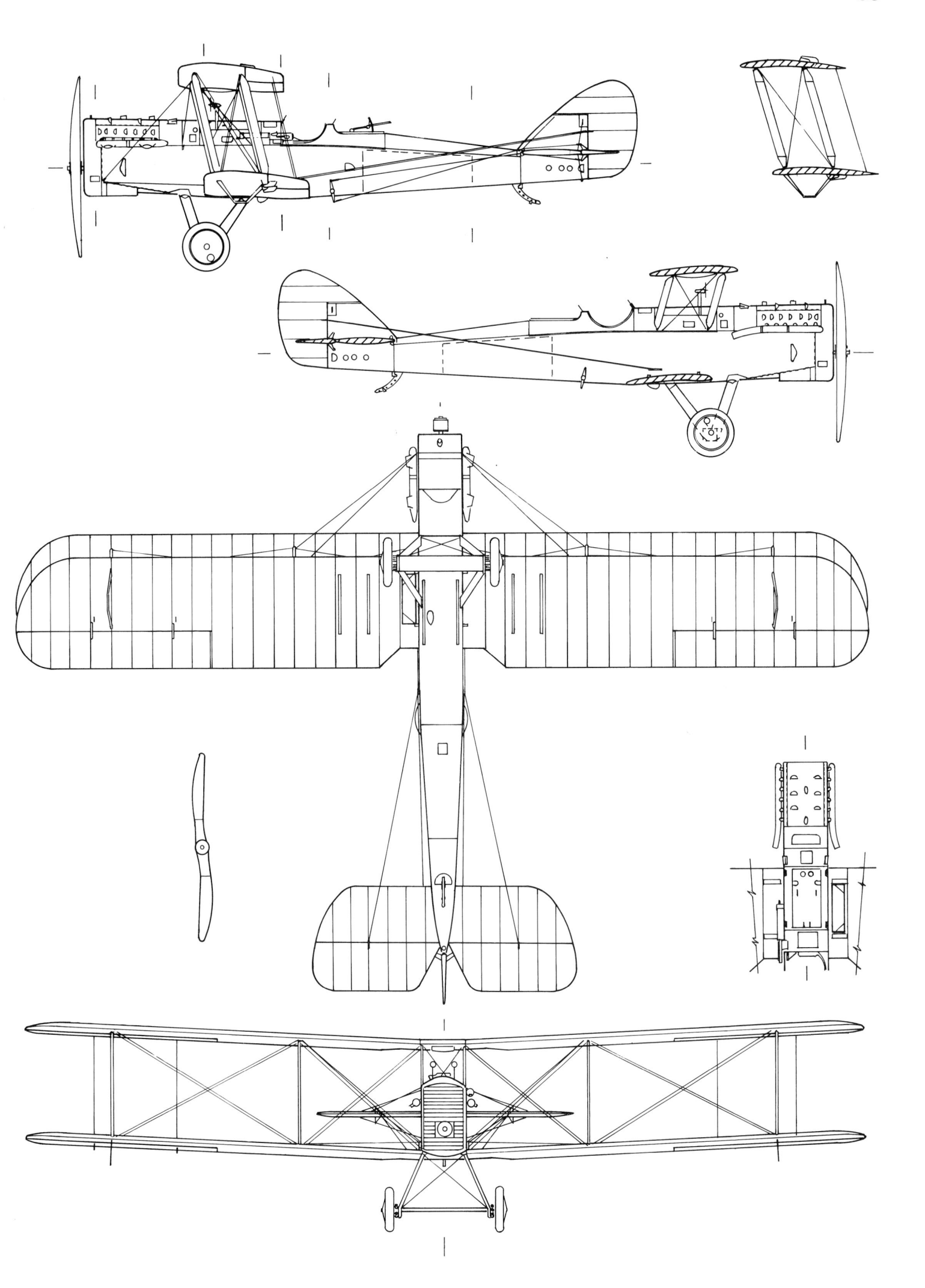

D.H.9A

D.H.15 Gazelle

Though basically a D.H.9A fitted with the 500 hp B.H.P. Atlantic V-12 water-cooled engine in place of the standard Liberty, this version of the design was given the designation D.H.15 and allotted the type name Gazelle.

The Atlantic engine comprised two 230 hp B.H.P. engines mounted on a common crankcase. Long exhaust pipes were fitted together with a large frontal radiator. In all other respects it was a standard D.H.9A airframe with the same armament. One aircraft was ordered on contract 35a/3015/C3446 on 7 September 1918.

J1937 was completed in July 1919. It carried out extensive trials of the Atlantic engine during 1919–1920 and was recorded at the ACE, Martlesham, in May 1920.

The end of the war ended plans to use the D.H.15 in the Independent Air Force in 1919.

The long exhaust pipe is the only external feature which distinguishes the D.H.15 from the D.H.9A. (Credit: J.M. Bruce/G.S. Leslie)

D.H.14 Okapi

Very much a stretched D.H.9A, the de Havilland D.H.14 was named Okapi after the African ruminant related to the giraffe but which is smaller with a short neck – which makes one wonder if there were people with a sense of humor in the Technical Department of the Air Department?

Designed as a successor to the 9A, the airframe was enlarged to accommodate a Rolls-Royce Condor engine. As it was thought that Condors would not be available the aircraft was to be engined with the Galloway Atlantic. In the event it was to emerge with the 600 hp Condor.

Three of these large single engined biplanes were ordered on 28 October 1918, as Contract No. 35a/3365/C3909 from the Aircraft Manufacturing Co. who soon changed their name to Airco. The external similarity to the 9A was noticeable and although of conventional construction the design featured novel features. Wind driven fuel pumps were not reliable at low speed. Because of the depth of the fuselage of the D.H.14 it was possible to blank off the upper half of the main tank and use this as a gravity tank, the wind driven pumps being used in cruising flight.

The pilot's Vickers gun was situated in the top decking. Steel tube was used for the engine mount. The gunner had the standard twin Lewis guns on a Scarff mounting. The tailplane was supported by four steel tube struts from underneath, thus reducing the likelihood of him shooting the tailplane bracing wires as had happened on the earlier De Havilland bombers. A conventional stout wooden vee strut landing gear was sprung by rubber cord.

The Armistice and post-war austerity saw the death knell of the D.H.14. The unfinished military aircraft were taken over by the newly formed de Havilland Aircraft Co. and completed at Stag Lane. Only two were delivered as military aircraft. J1938 was completed on 22 September 1920, and had its first flight that month. It was at Farnborough in October where Sqn. Ldr. Roderic Hill used it for Condor engine trials. It underwent a leisurely test procedure at the RACE and the ACE. On 10 February 1922 it forced landed at Burnham Beeches, hit the top of a tree, crashed, and burned out.

The second aircraft, J1939, was completed in December 1920 and at Cricklewood the following January. After evaluation at the ACE in March it was re-engined with a Condor Ia in November. Although one was being flown in civil markings the D.H.14 remained on the Secret List until one was statically exhibited at the Imperial Air Conference at Croydon in February 1922. After being damaged beyond repair J1939 was written off the following April.

The third D.H.14, J1940, was completed first as a two-seat mail plane. F.S. Cotton proposed to use the aircraft to compete for the Australian Government's prize of £10,000 for the first flight from the UK to Australia. The Vickers Vimy of the Smith brothers won the prize before Cotton's aircraft was ready. Registered G-EAPY and completed as a D.H.14a with 450 hp Napier Lion engine, the aircraft was fitted with a special four-wheeled undercarriage. Cotton now attempted to fly London to Cape Town but came to grief when he had a forced landing while looking for a non-existent aerodrome in Messina, Italy. The aircraft was shipped back to the UK and repaired. Cotton competed in it in the 1920 Aerial Derby but it was damaged beyond repair in a forced landing at Hetford.

The sheer size of the D.H.14 is evident in this view. (Credit: J.M. Bruce/G.S. Leslie)

D.H.14
(Plan)

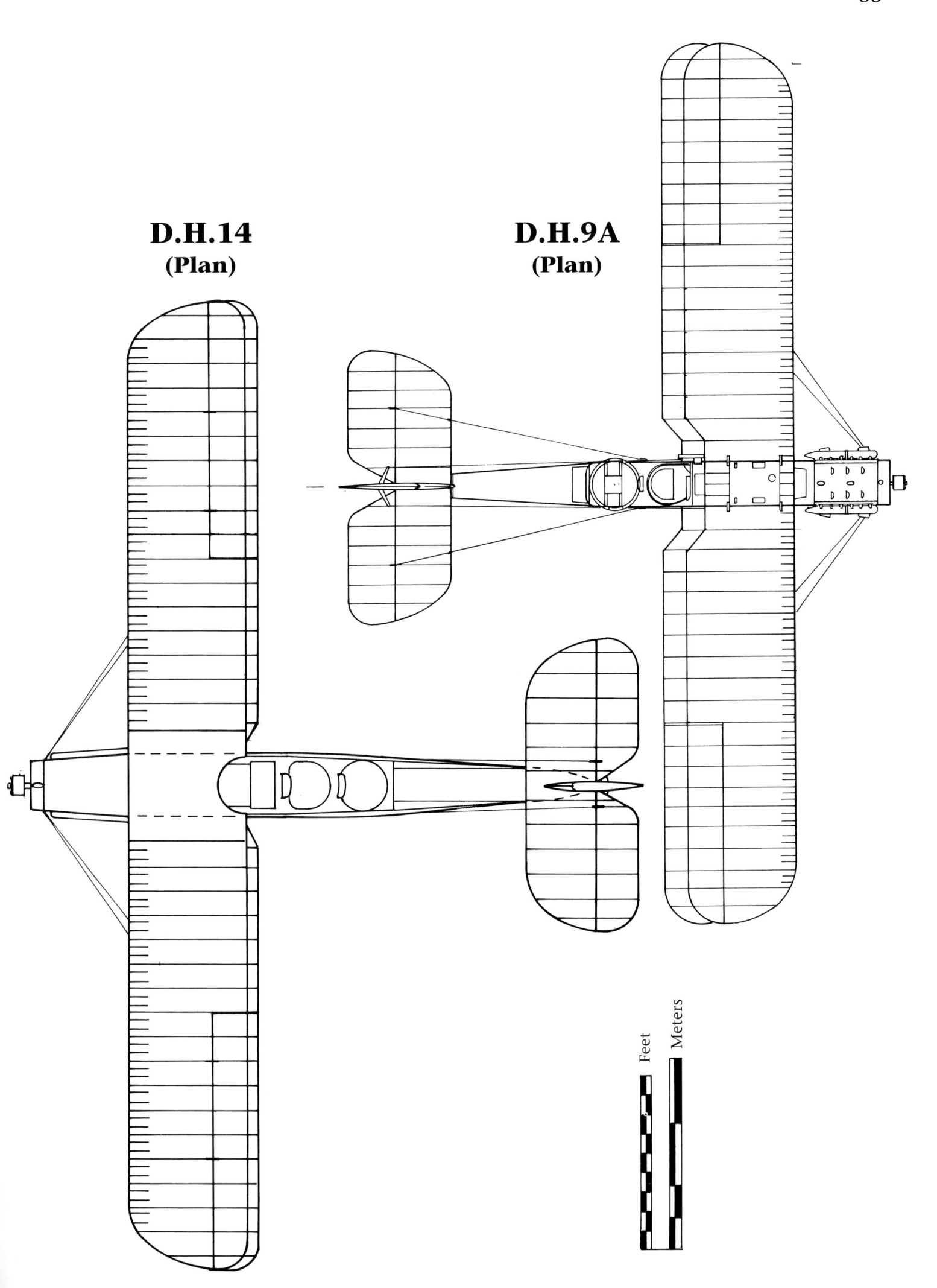

D.H.9A
(Plan)

55

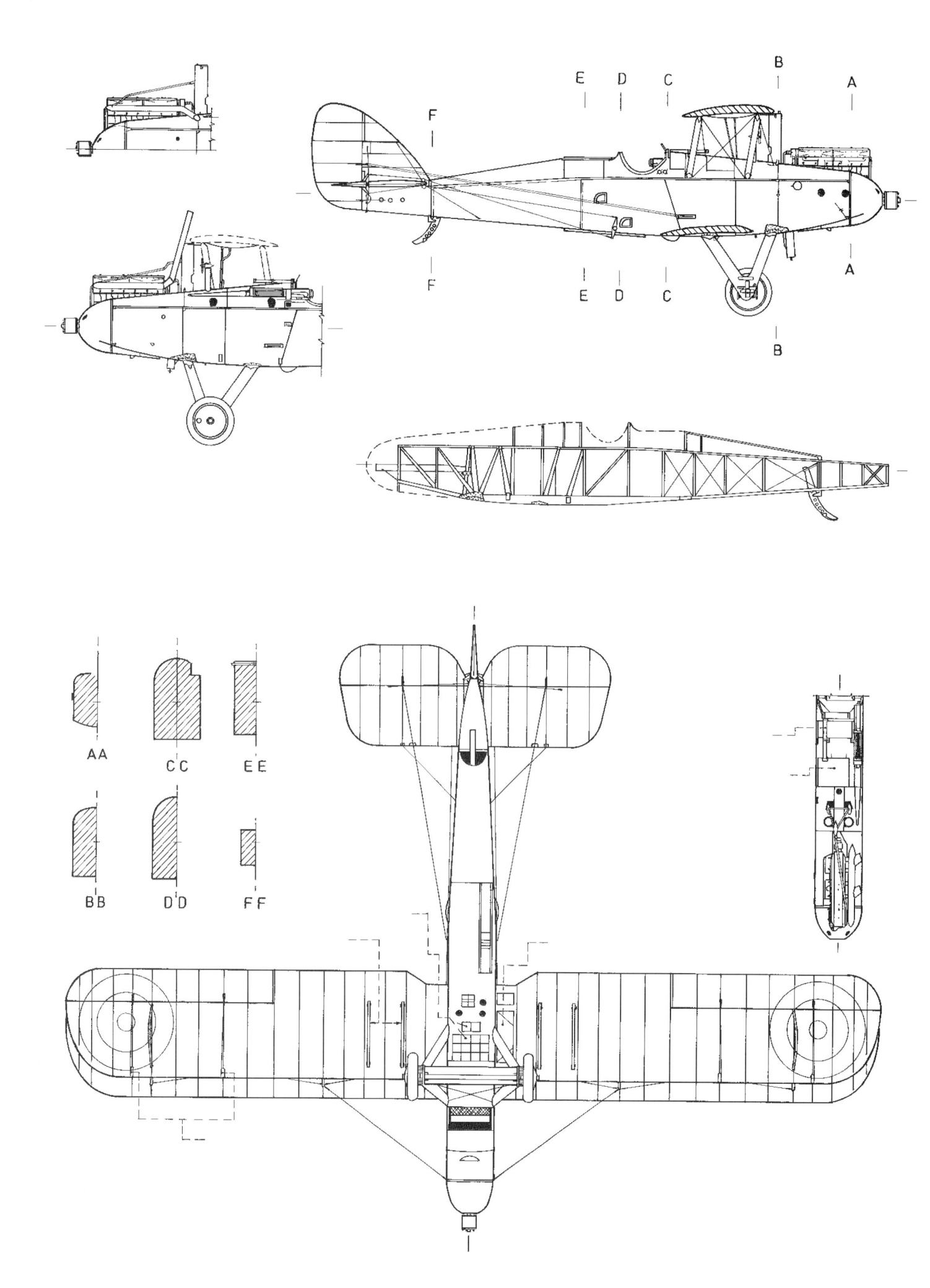

D.H.9

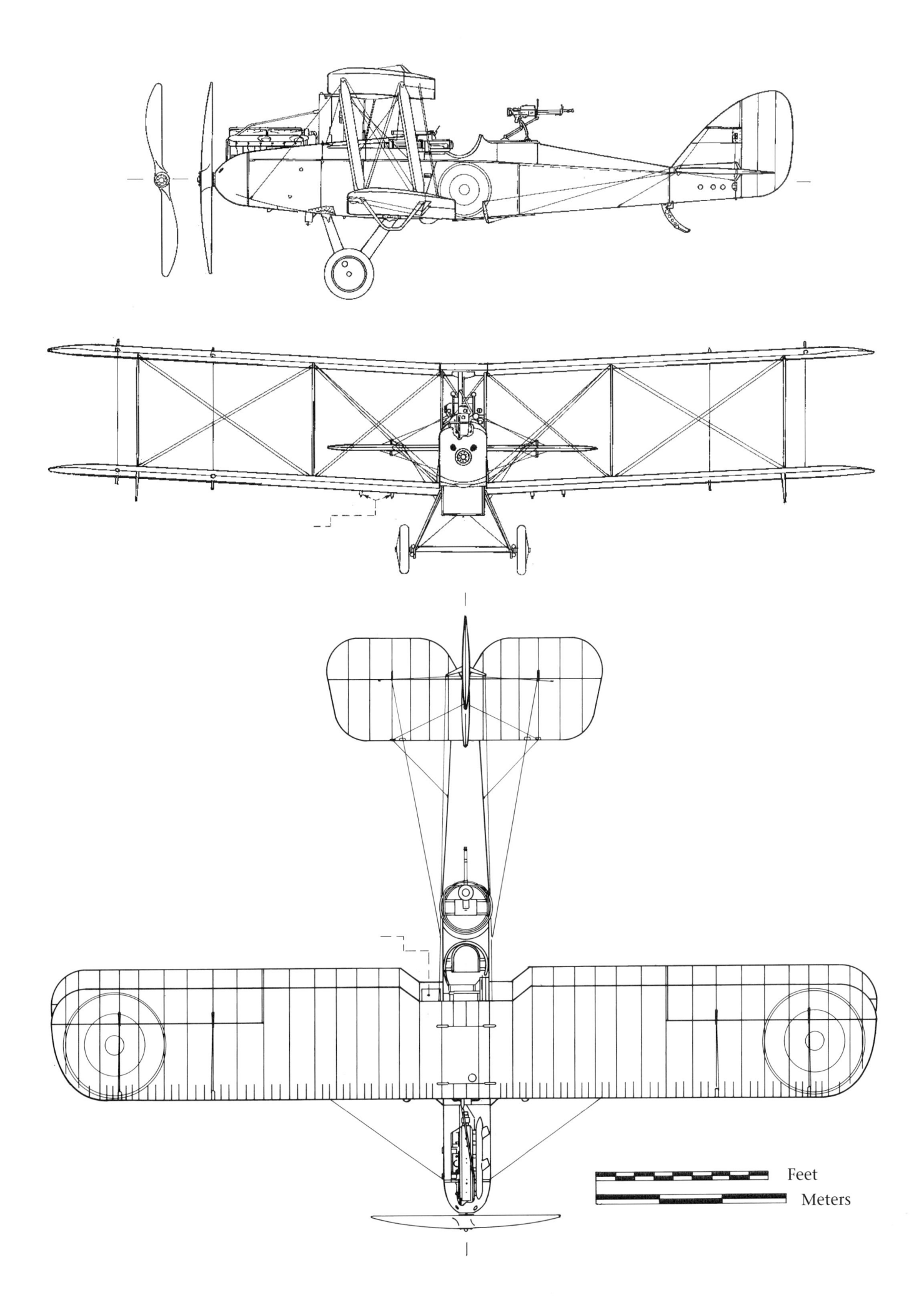

D.H.9

Feet
Meters

D.H.10

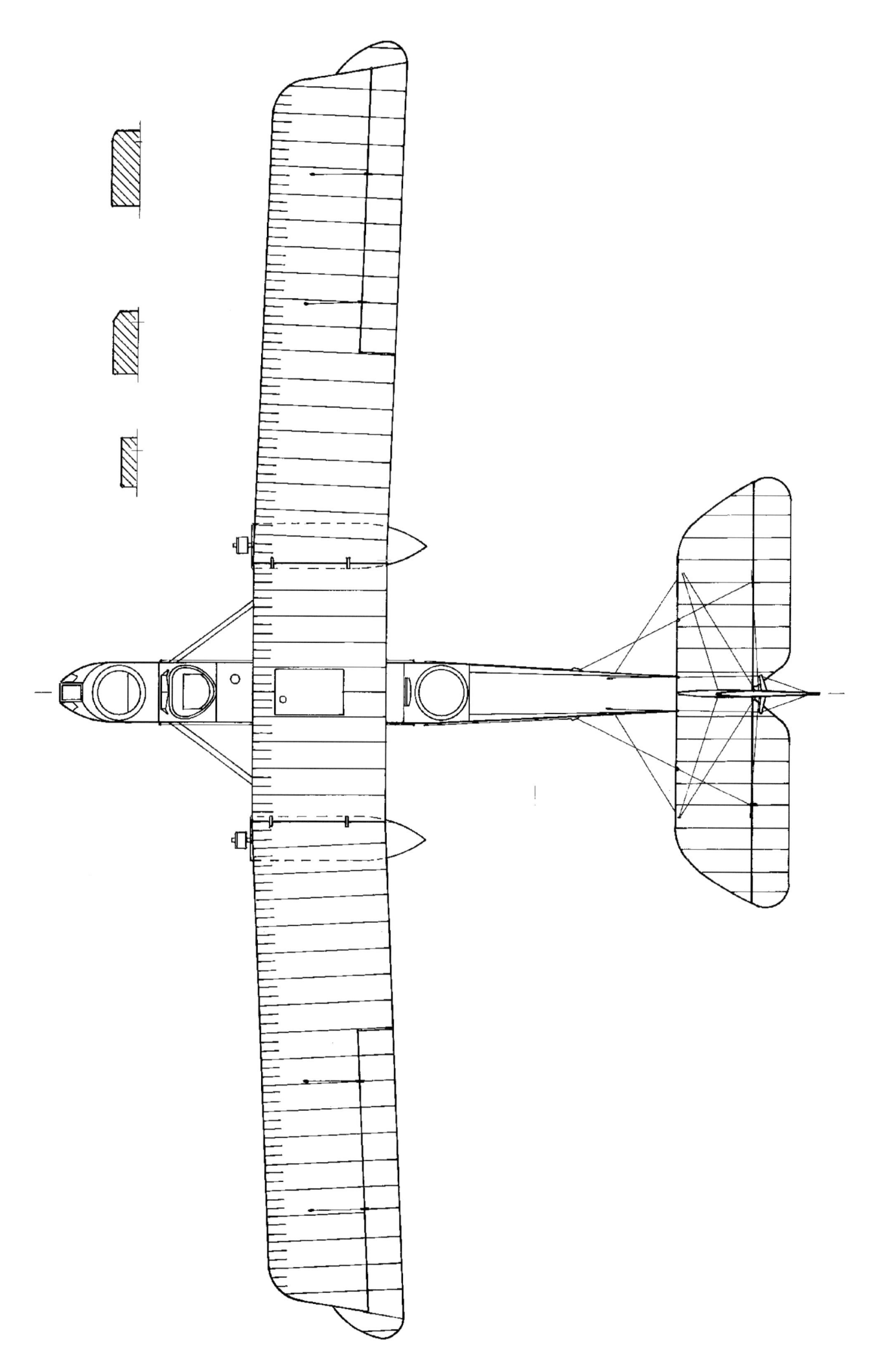

D.H.10

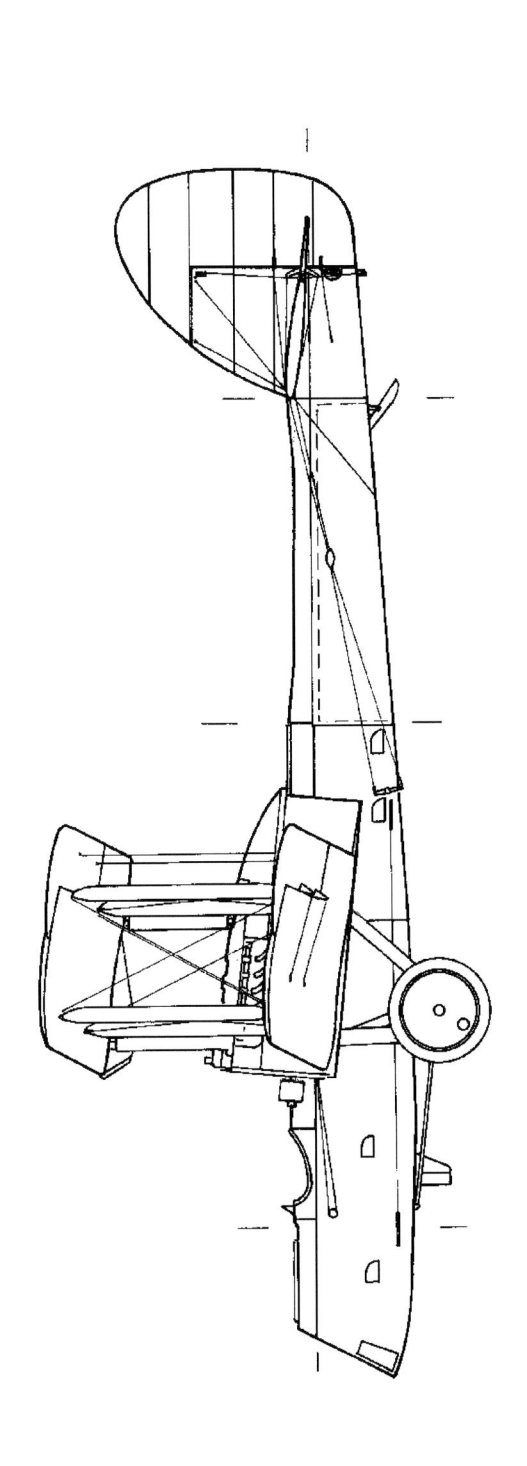

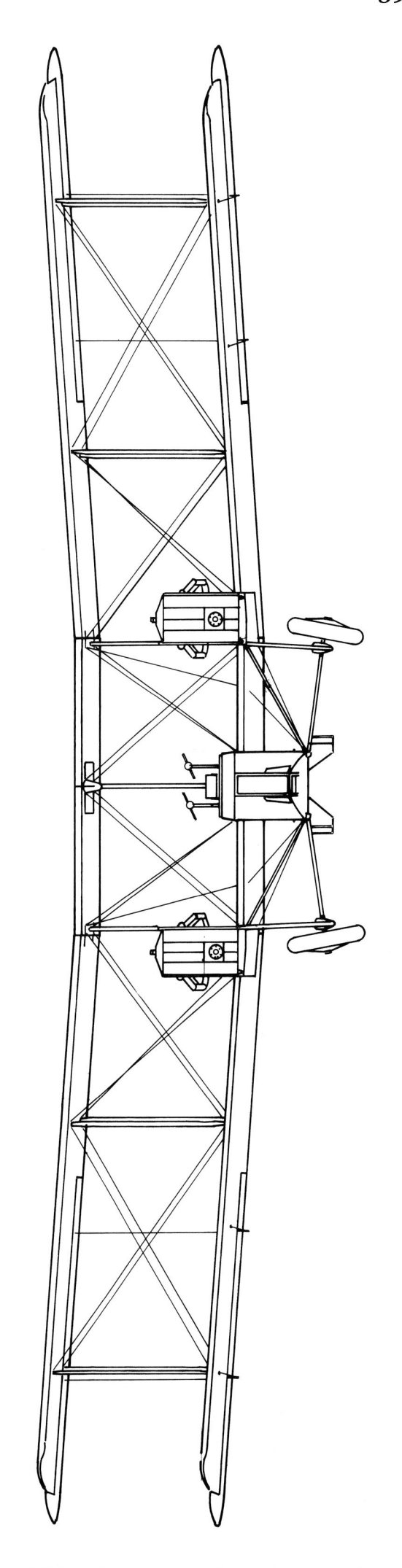

D.H.11

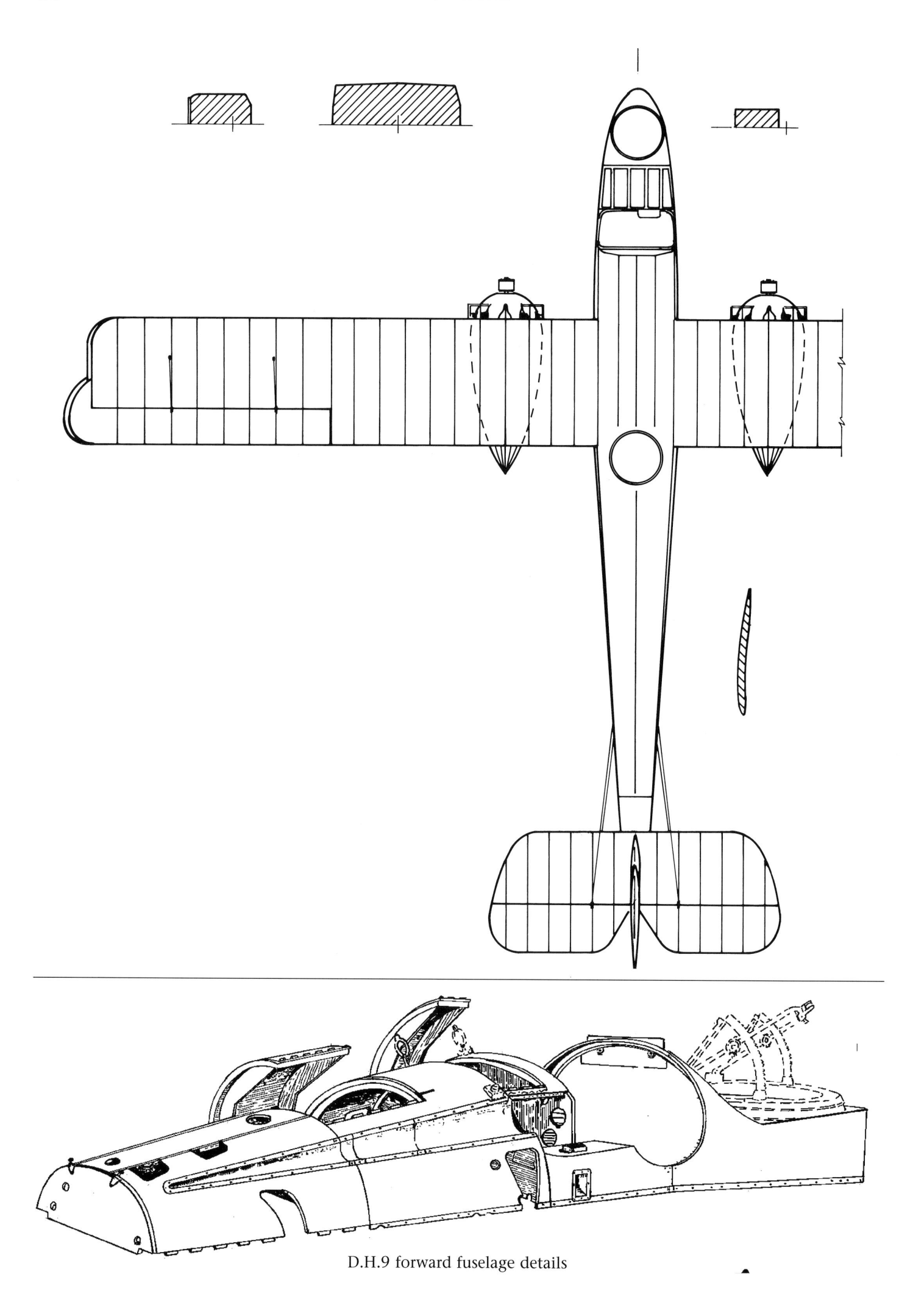

D.H.9 forward fuselage details

D.H.11

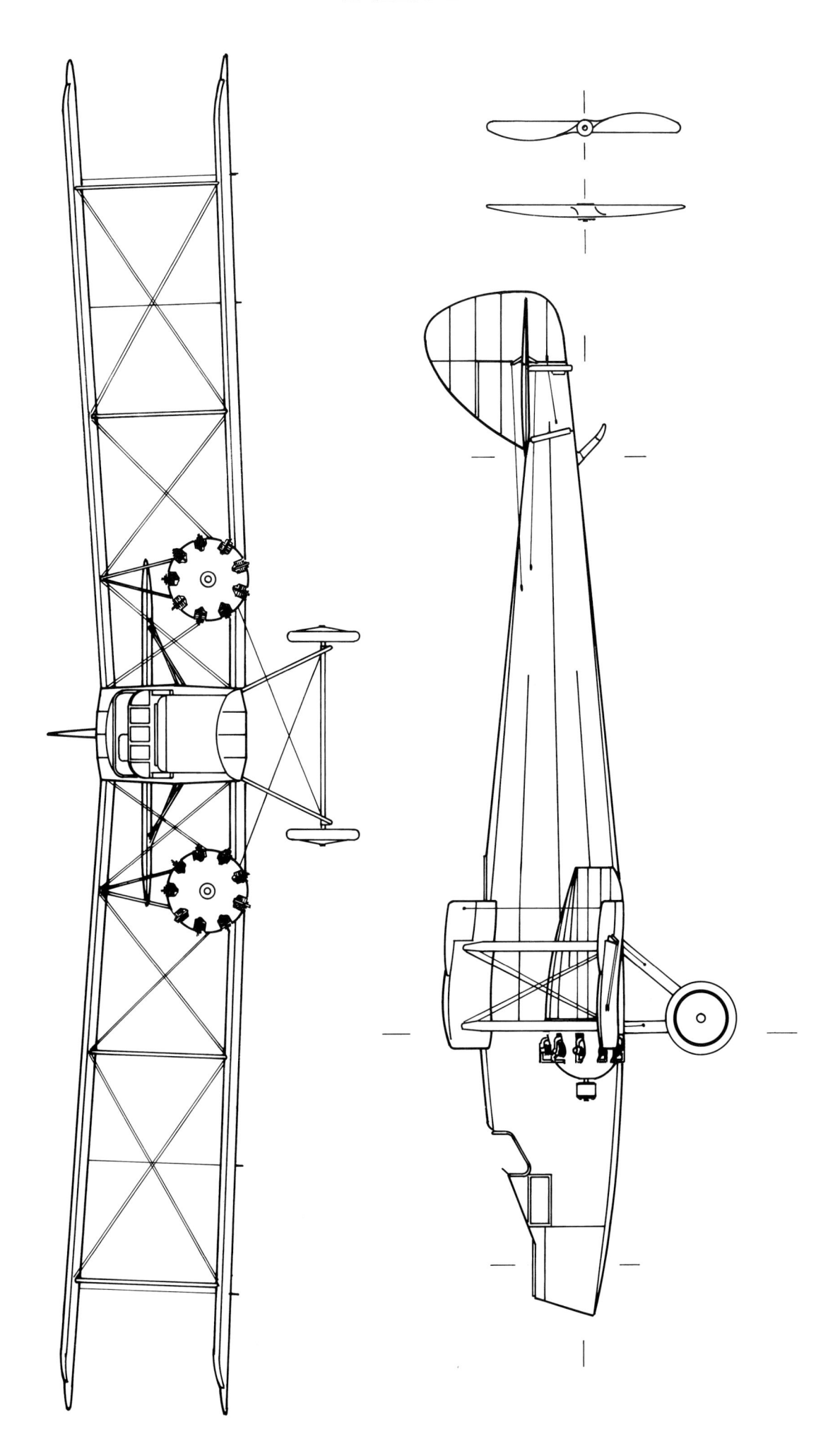

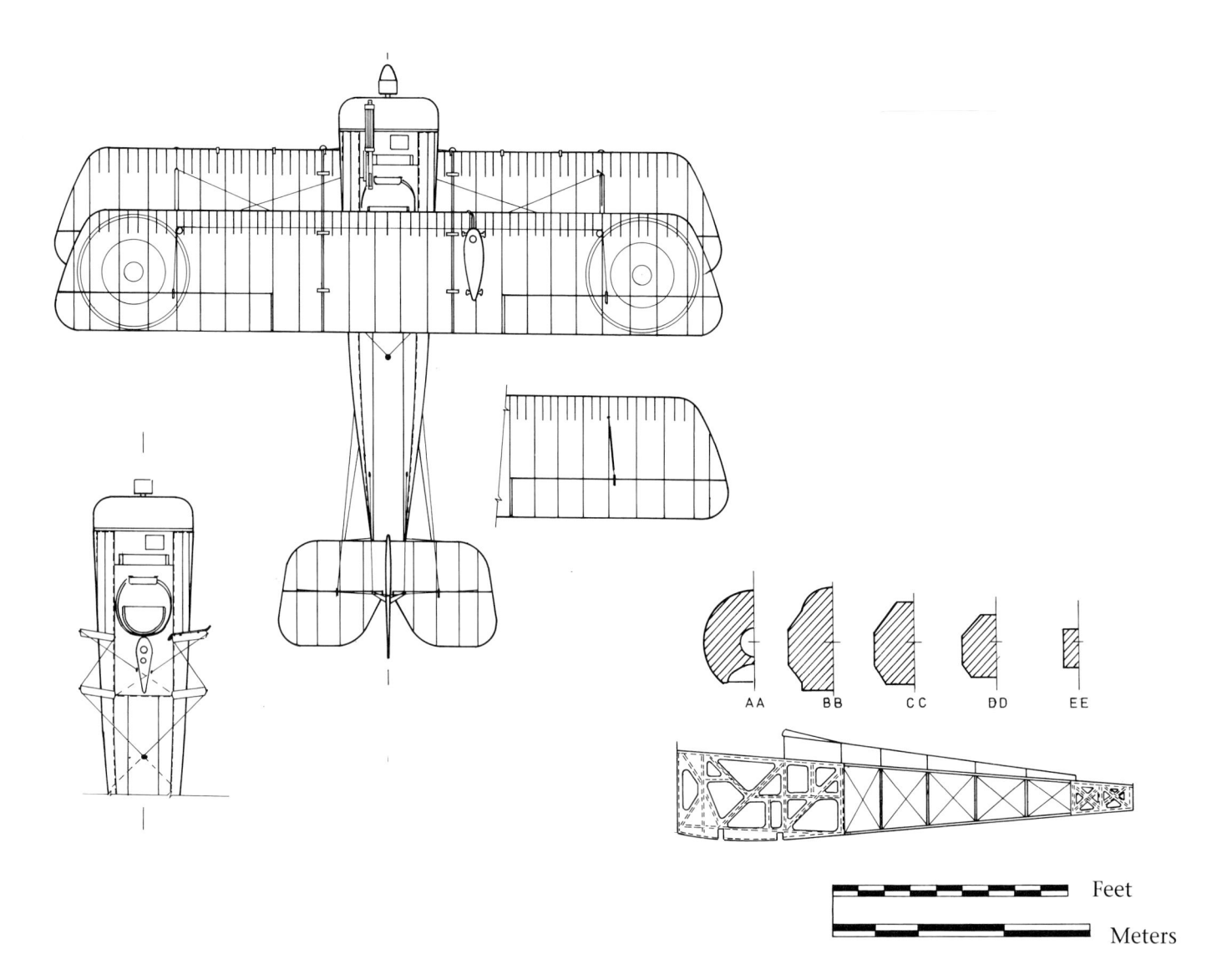

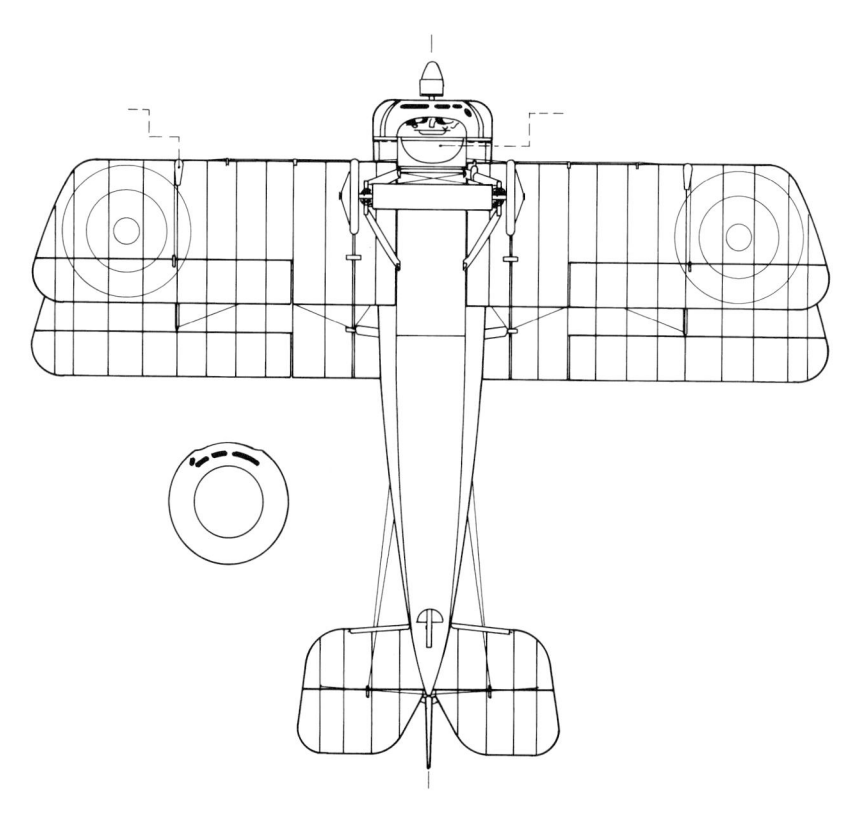

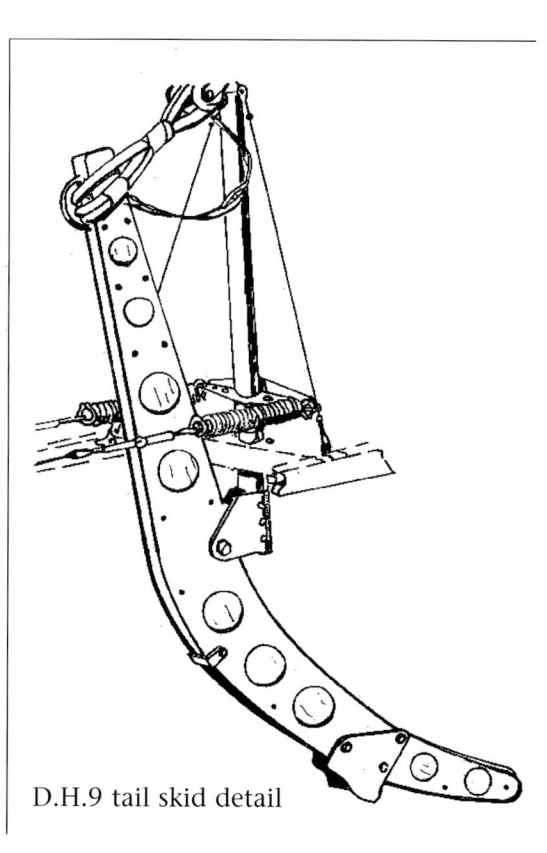

D.H.5

D.H.9 tail skid detail

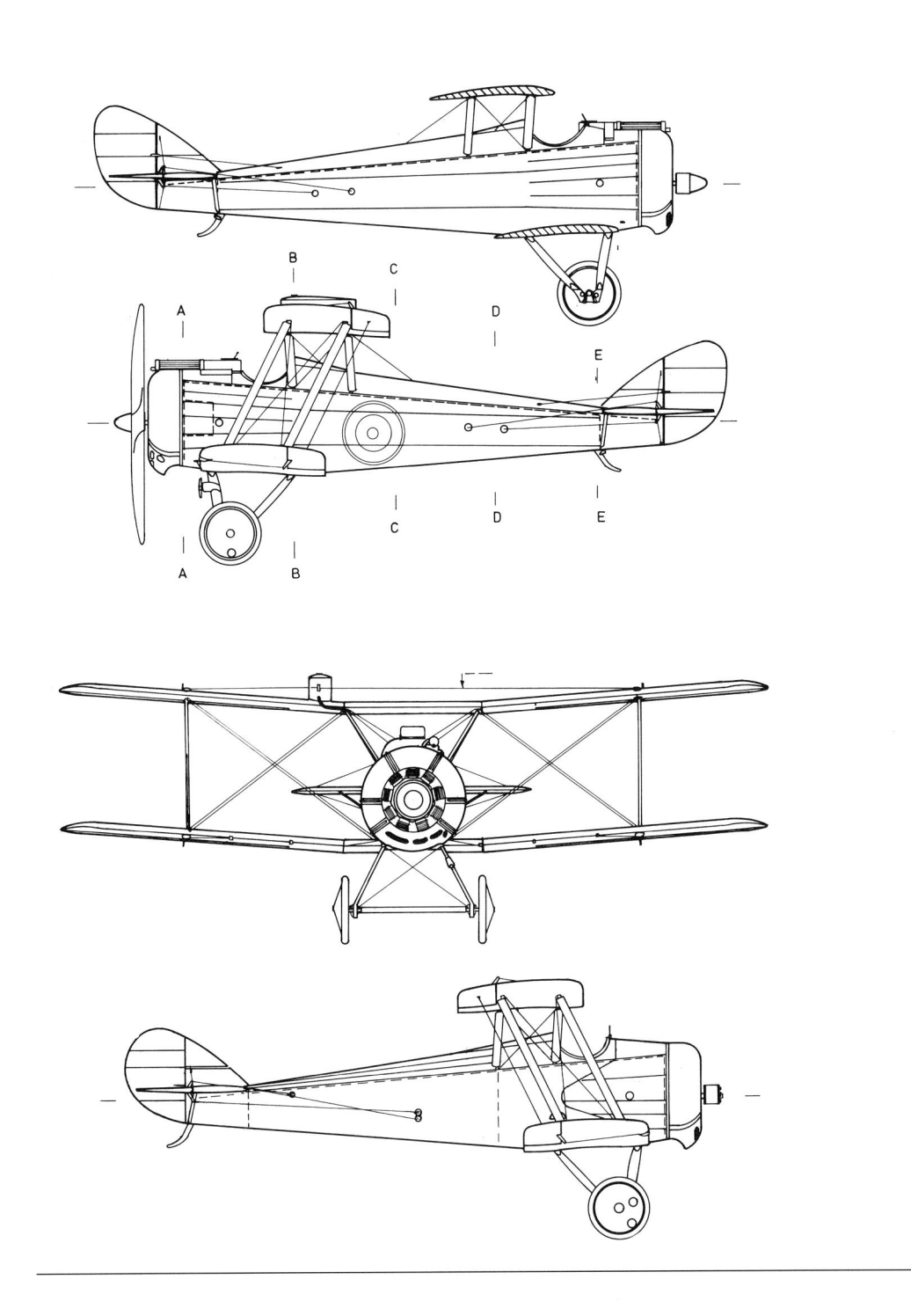

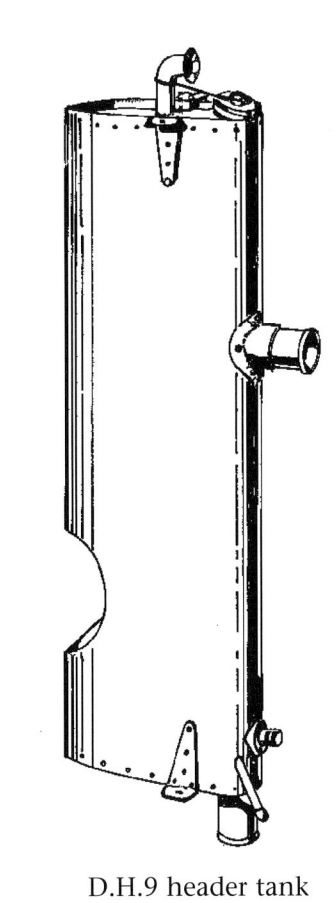

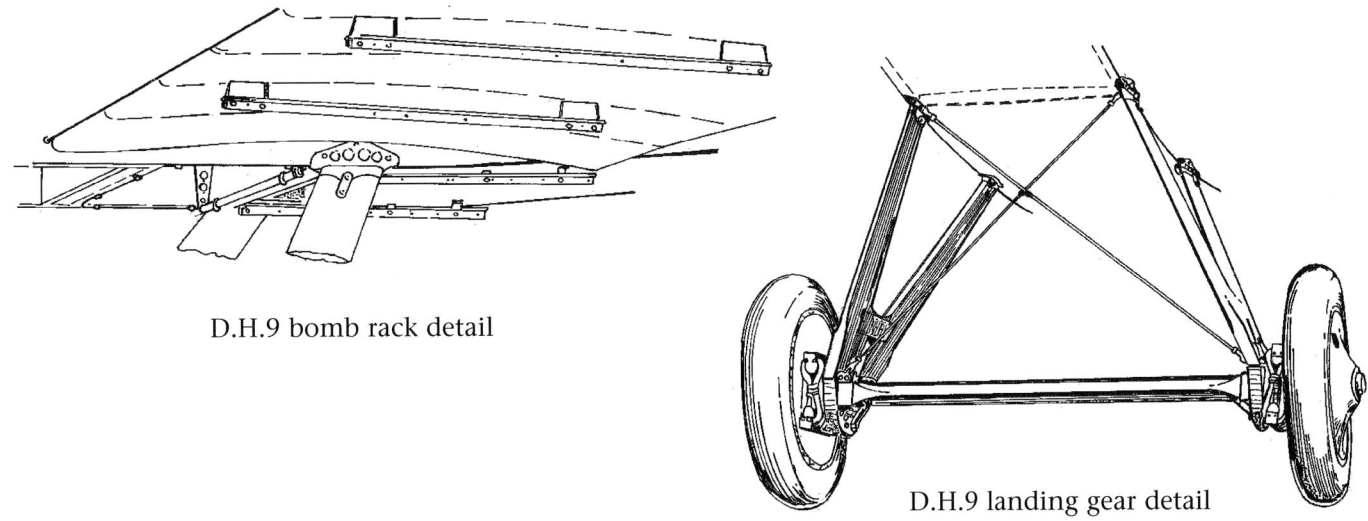

D.H.5

D.H.9 header tank

D.H.9 bomb rack detail

D.H.9 landing gear detail

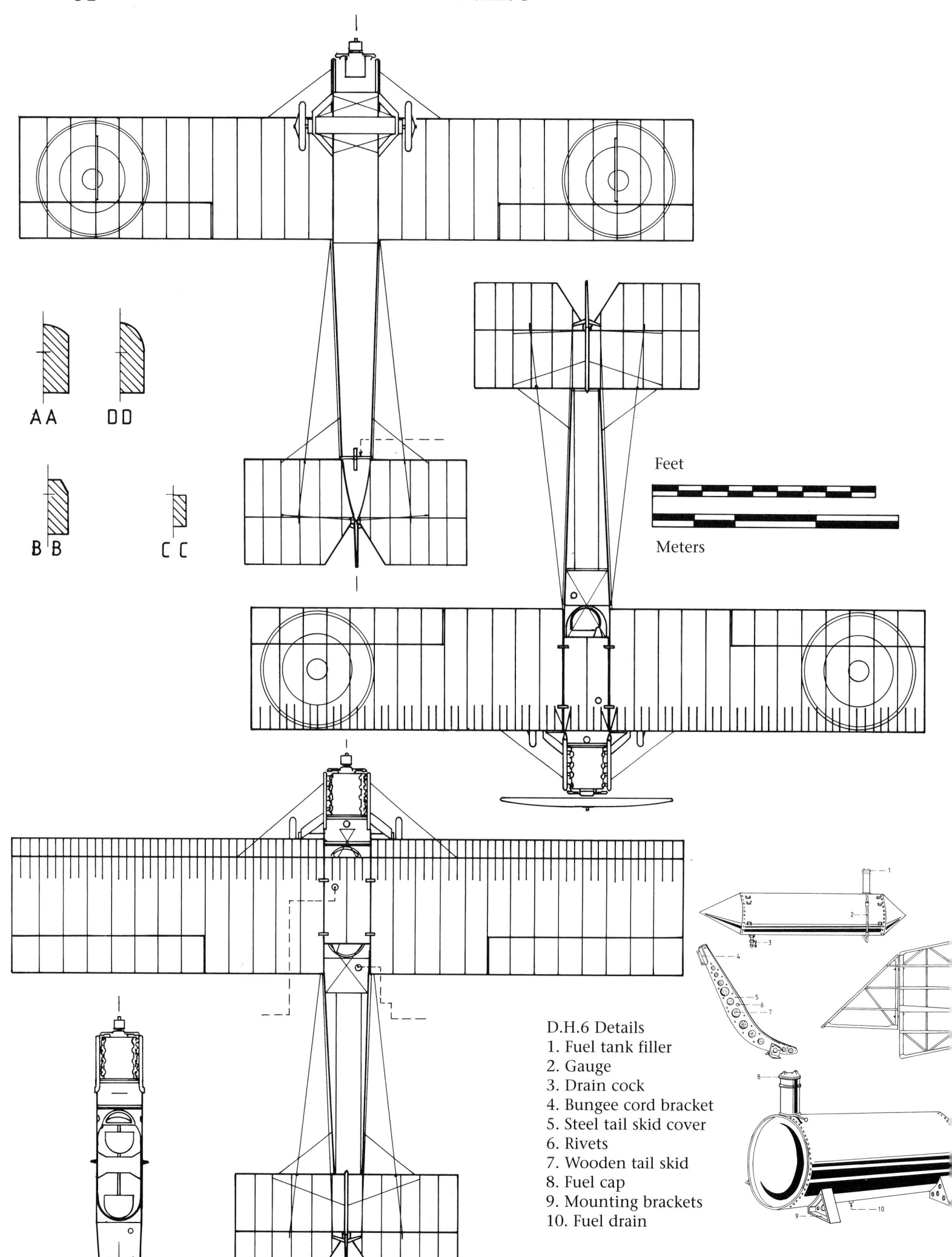

D.H.6

D.H.6 Details
1. Fuel tank filler
2. Gauge
3. Drain cock
4. Bungee cord bracket
5. Steel tail skid cover
6. Rivets
7. Wooden tail skid
8. Fuel cap
9. Mounting brackets
10. Fuel drain

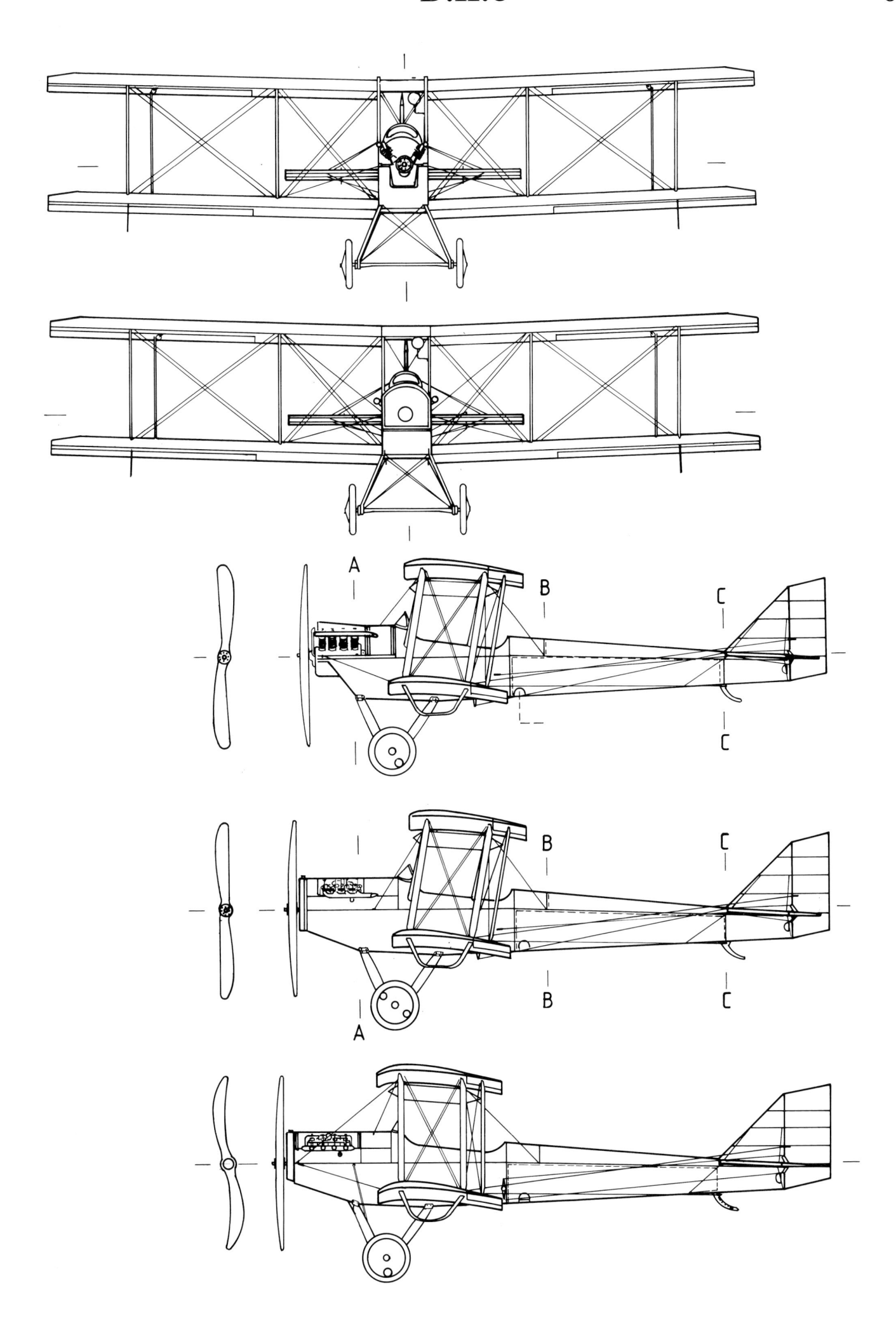

D.H.6

D.H.6

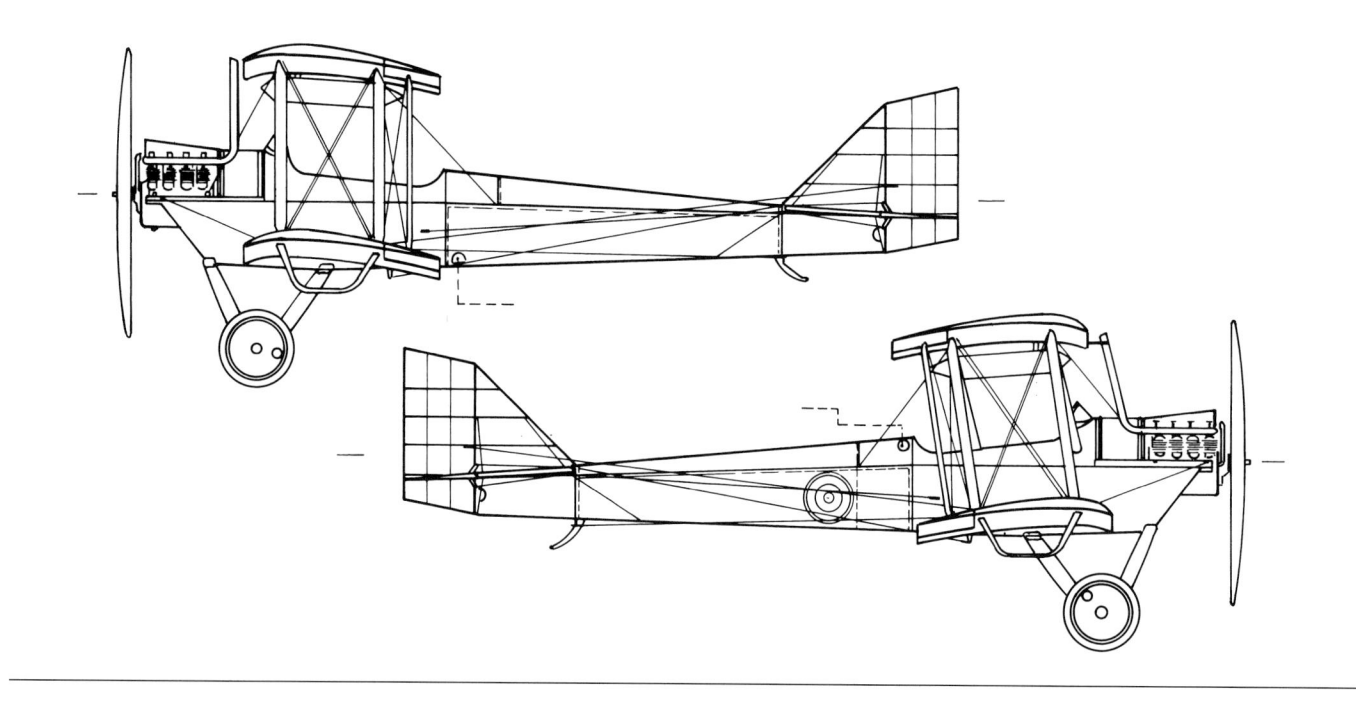

D.H.5

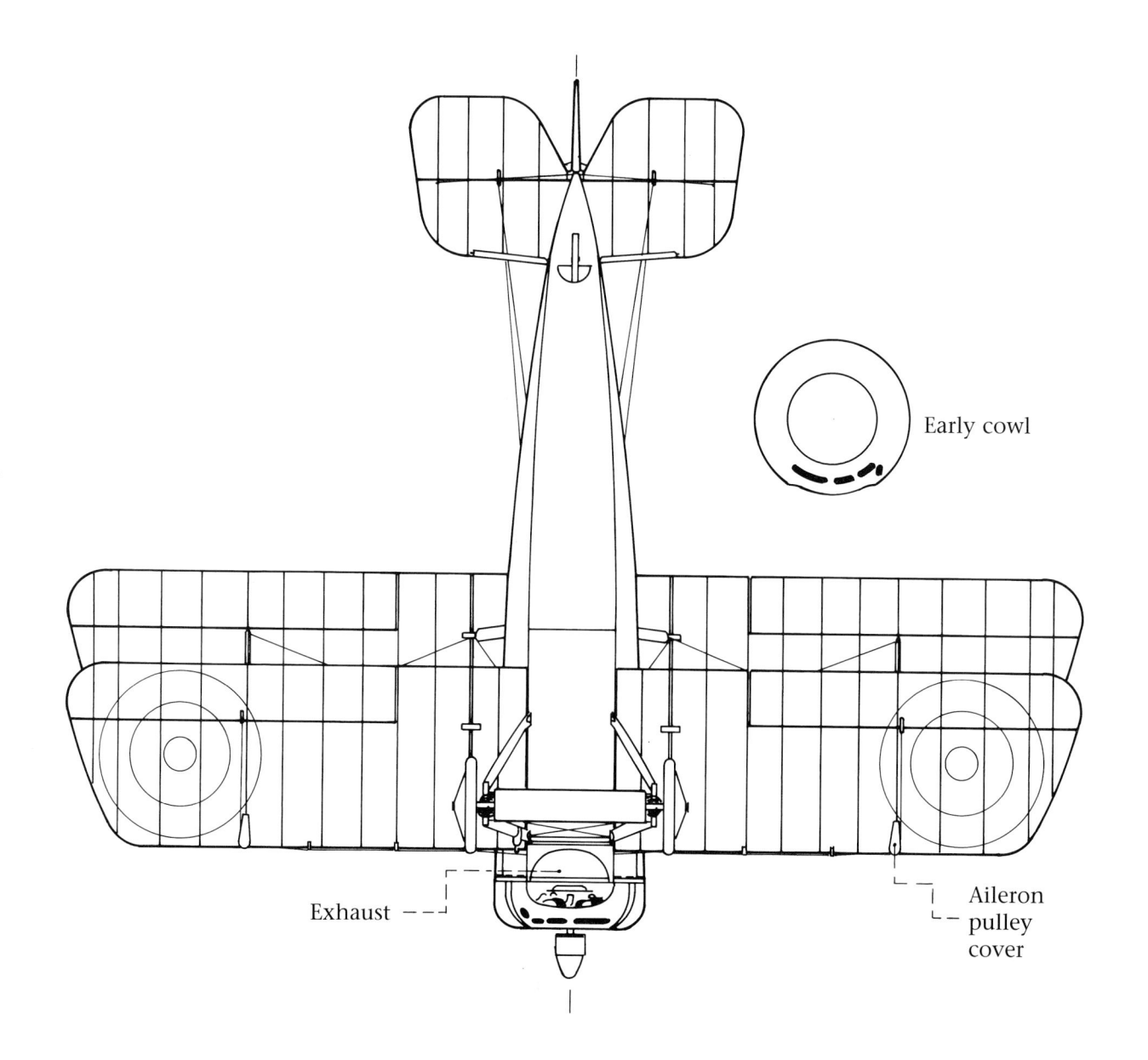

Early cowl

Exhaust

Aileron pulley cover

D.H.5

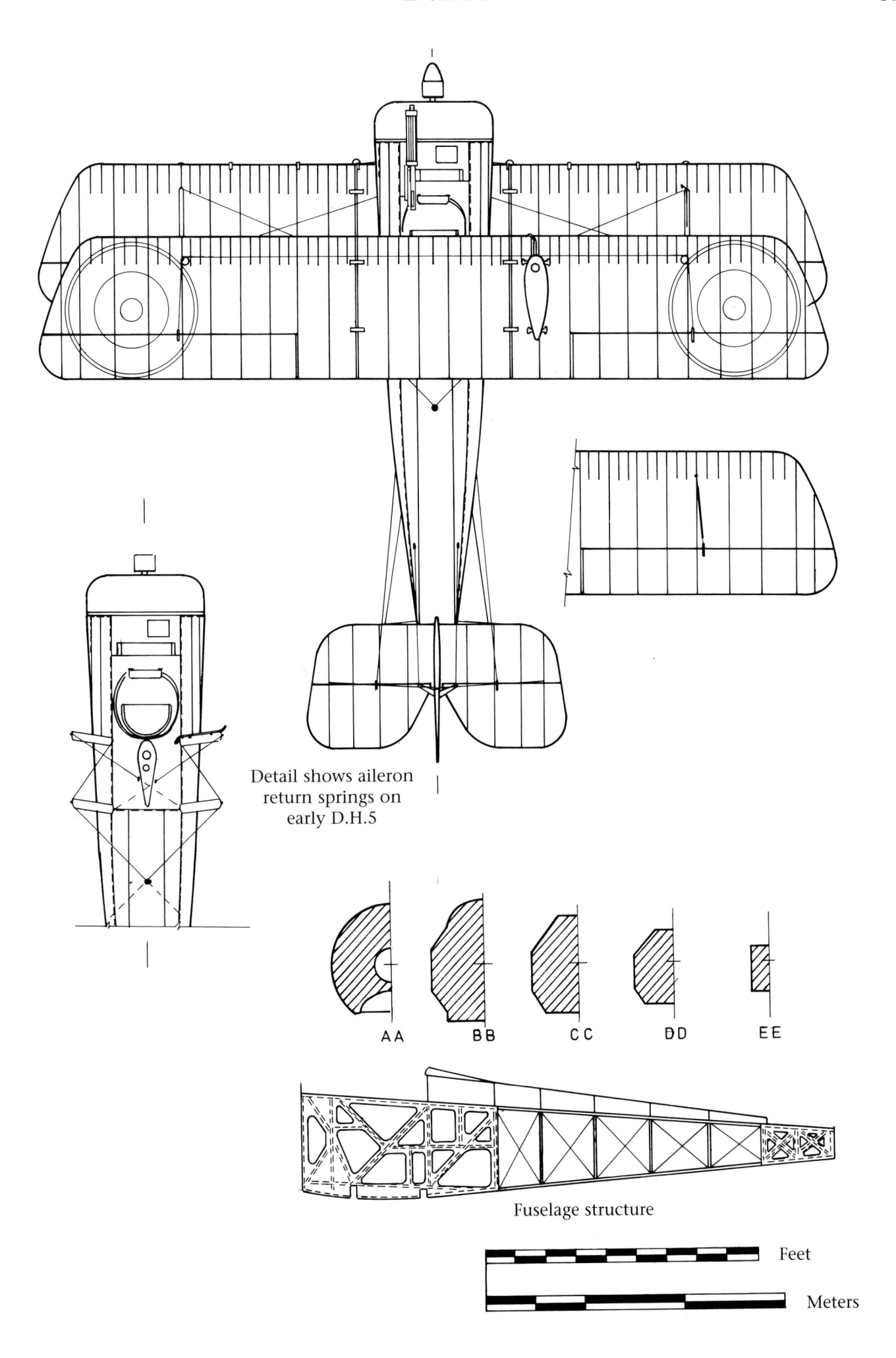

Detail shows aileron return springs on early D.H.5

Fuselage structure

Feet

Meters

D.H.5

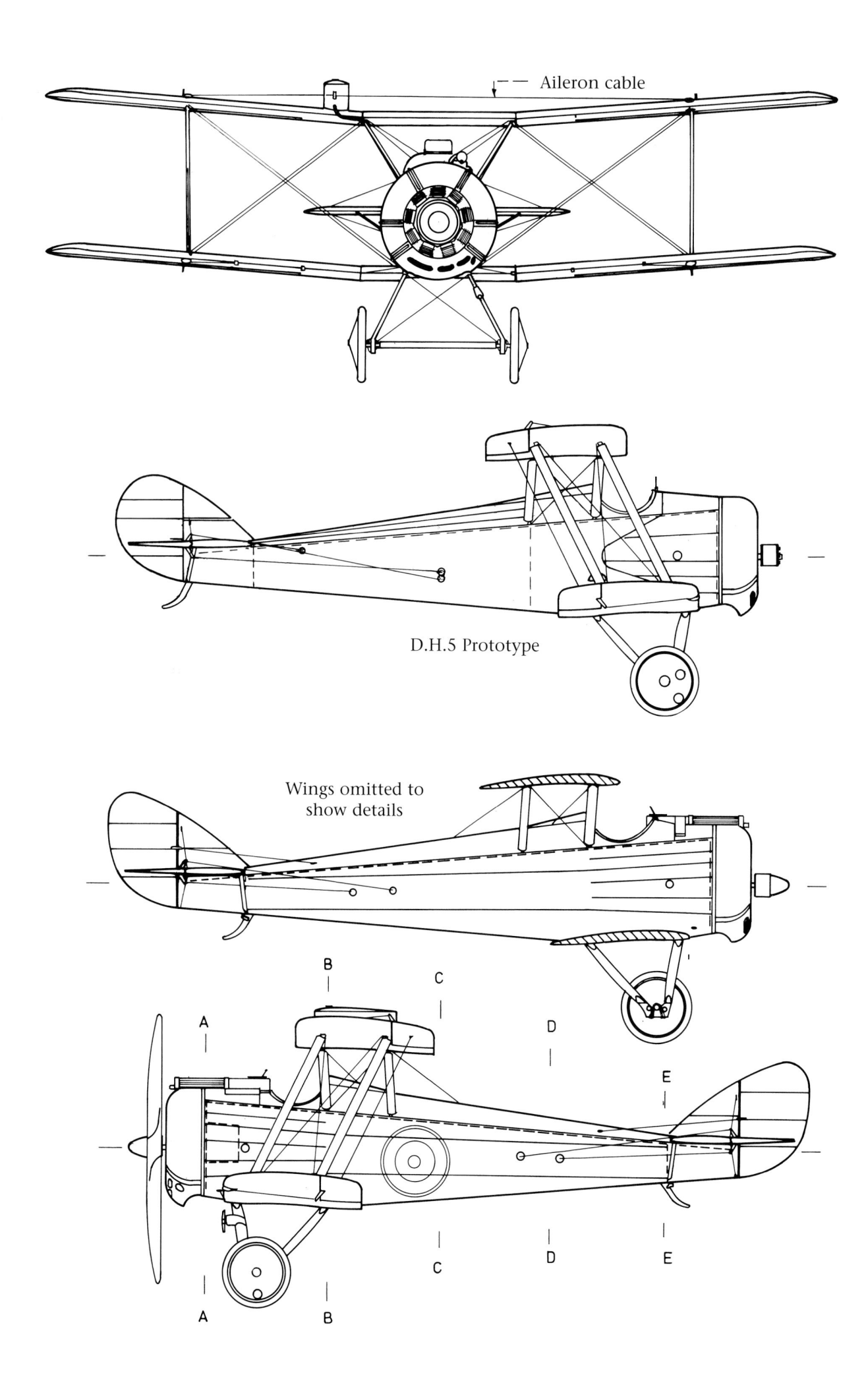

D.H.5 Prototype

Wings omitted to show details

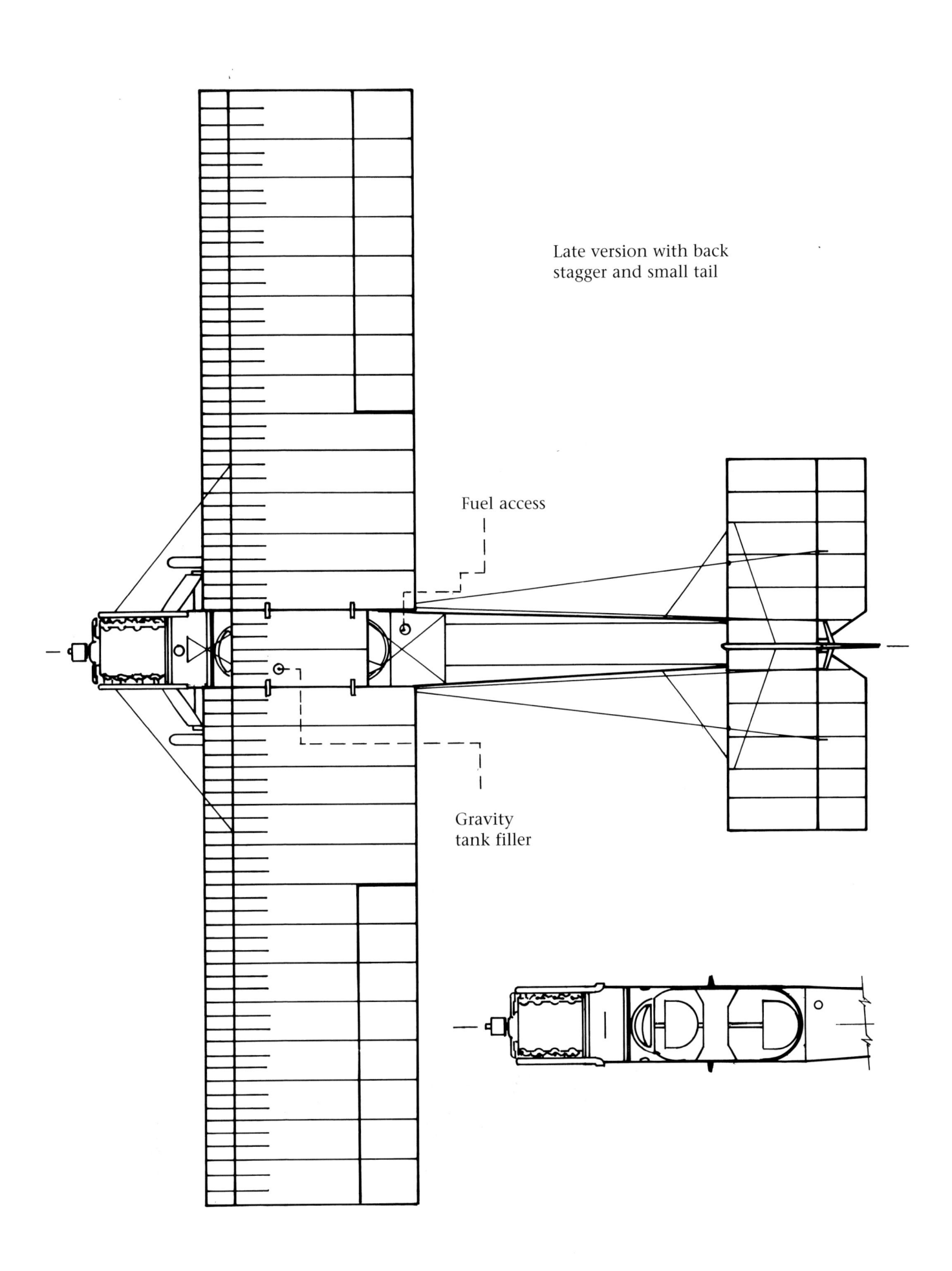

D.H.6

Late version with back stagger and small tail

Fuel access

Gravity tank filler

D.H.6

Plan view: Early version without back stagger and large tail

Feet
Meters

Side view: Early production without backstagger, RAF 1a

Footstep

D.H.6

71

Bottom view: Early version without back stagger and large tail

Side view: Early production with backstagger, RAF 1a

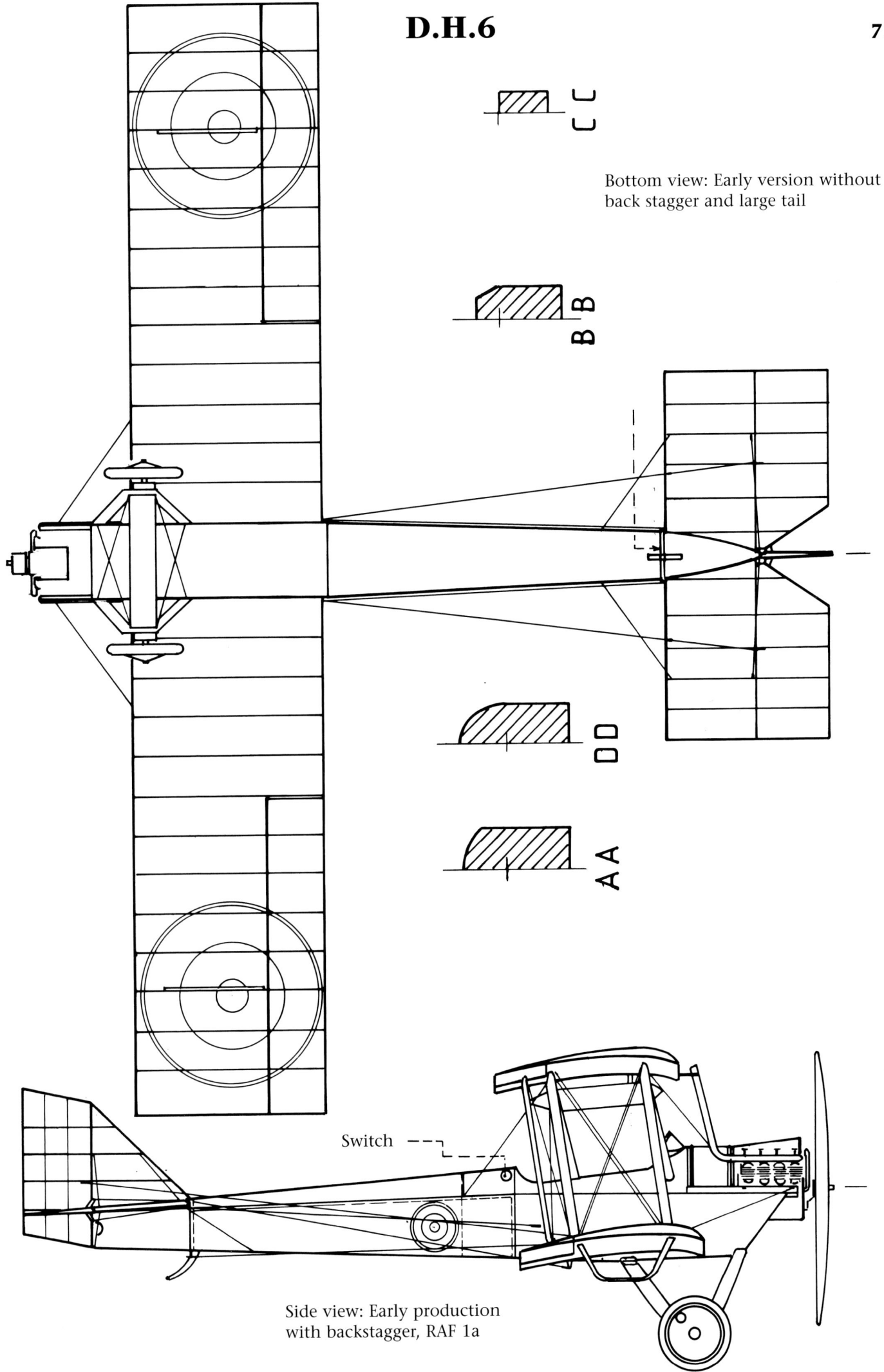

Switch

D.H.6

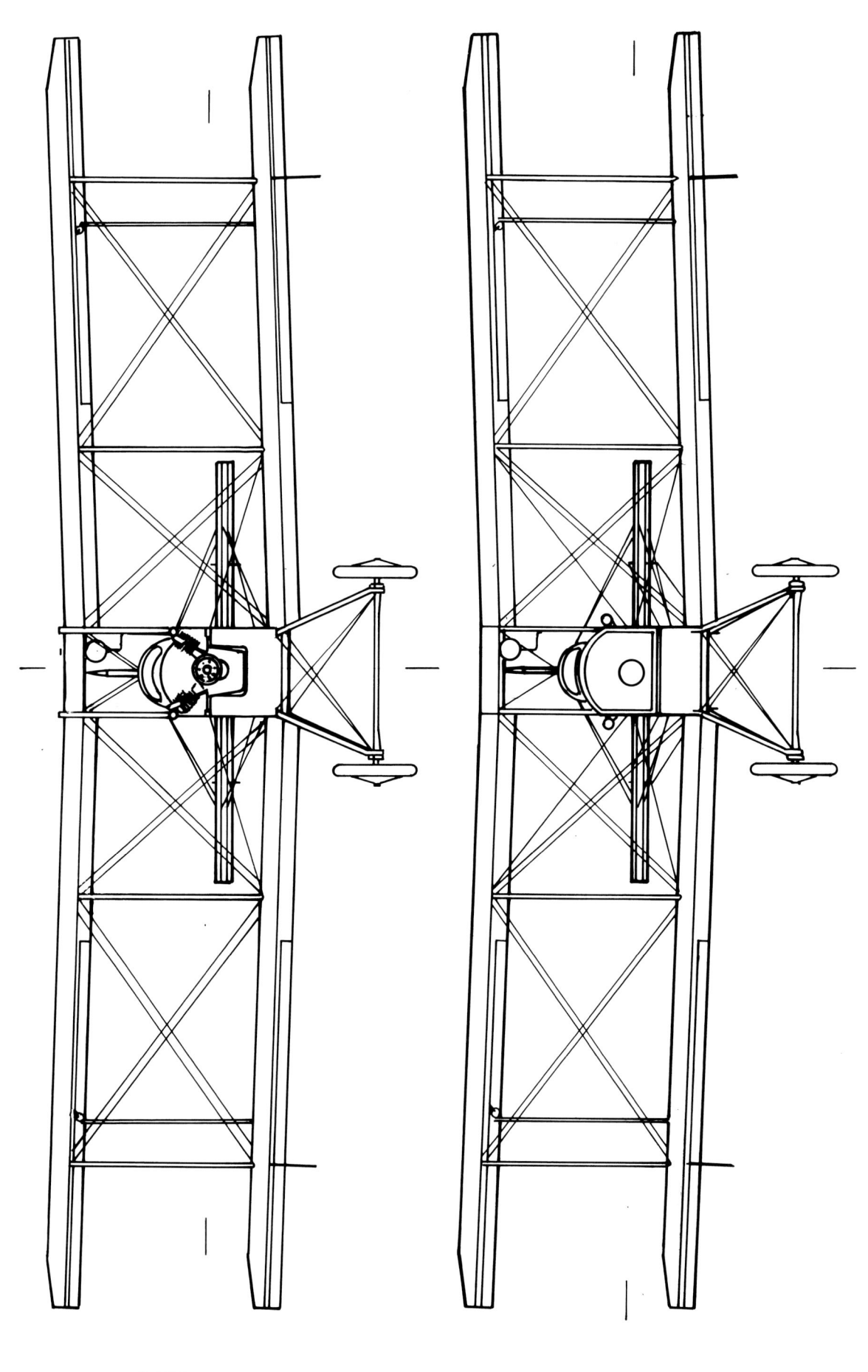

RAF 1a   OX-5

D.H.6

Renault (late tail)

OX-5 (late tail)

OX-5 without backstagger

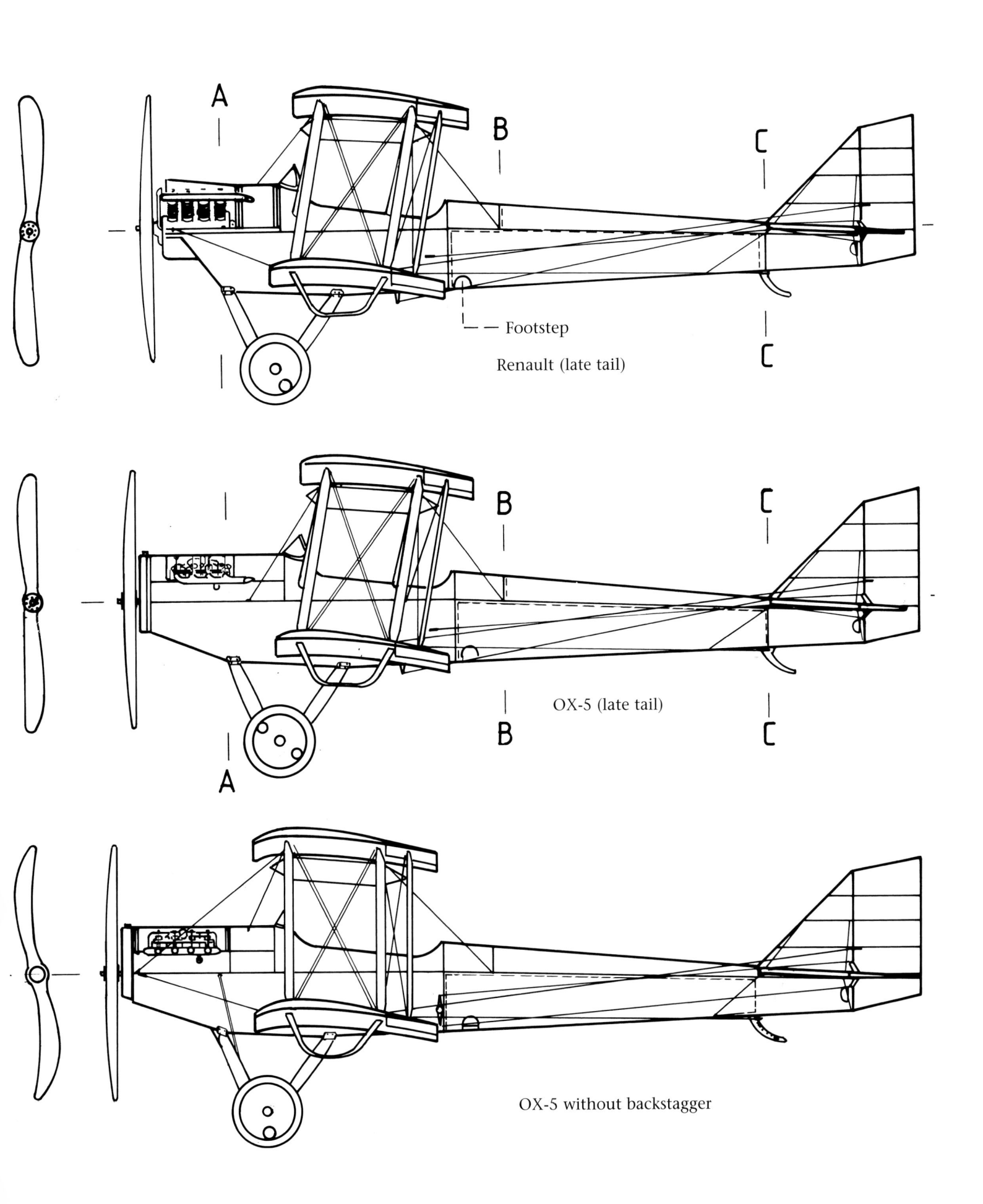

Colors and Markings

Early coloring was introduced by the British as a means of preserving fabric. Fabric was doped and then varnished. V.114 was clear varnish while P.C.10 was the same as V.114, except that it contained a "Protective Coloring." The early de Havillands (D.H.1, D.H.2, and D.H.4) were finished in clear doped fabric surfaces with the metal and ply panels painted grey. As P.C.10 was introduced it was applied to the upper surfaces of wings, tailplanes, fin, and fuselage fabric panels. Ply and metal panels were also finished in a matching color. Under surfaces of wings were left clear doped as were under fuselage fabric areas.

P.C.10 was a khaki color and it was followed by P.C.12, a reddish brown color. P.C.12 was for use in hot climates, such as the Middle East. Lawrence Wackett, who had experience with No.1 Squadron, AFC, considered P.C.12 to be a far superior scheme than the V.84 aluminium dope introduced after the war. The RNAS seems to have a preference for P.C.12 over P.C.10 and some of its D.H.4 biplanes may have carried this coloring. Material shortages would have seen best use made of which ever dopes and finishes were available.

Post-war the D.H.9A and D.H.10 bombers used by the British in the Middle East and the United kingdom were finished in V.84 aluminium.

The interpretation of monochrome photographs depends on the lighting, the age and use of the aircraft, the time period, and how many times the photograph has been copied. The best available data has been used in preparing the profile drawings.

De Havilland D.H.5

1. (Inside front cover) D.H.5 serial A9449, aircraft "1" of No. 68 (Australian) Squadron. This is the aircraft on which Capt. G.C. Wilson won his MC during the Battle of Cambrai. P.C.10 overall with clear doped fabric lower surfaces. White individual and squadron markings. Note late cowl and that this aircraft had the revised aileron system.

2. D.H.5 serial A9242 of No. 68 (Australian) Squadron (later No. 2 Squadron, AFC). The profile depicts the aircraft as it appeared at Harlaxton, UK, while the squadron was working up on the D.H.5 fighter. One of a number of D.H.5 which were raised by patriotic subscription, this aircraft bears the inscription "New South Wales No. 14. Women's Battleplane. Subscribed and collected by women of New South Wales" in white under the cockpit. Band is also white. Rest of the airframe is P.C.10 overall with clear doped lower fabric surfaces. Wheel covers possibly blue. This aircraft went with the squadron to France. Note early cowl and Thorton Pickard camera gun. The revised aileron control that replaced the bungee return springs has been fitted. It is not known if it retained these markings; if so then they would have been supplemented by an individual aircraft number and squadron markings.

3. D.H.5 serial A9474, aircraft "F" of No. 41 Squadron. This aircraft was issued to the squadron on 14 October and was shot down on 29th, the pilot, 2/Lt. F. Clark, becoming a POW. P.C.10 overall with clear doped fabric lower surfaces. White individual and squadron markings. White outline to rudder serial.

4. D.H.5 serial A9340, Aircraft "C" of No. 32 Squadron, RFC. This aircraft served with the squadron from 24 October 1917 until March 1918. It was then displayed as part of a campaign to raise funds for the war effort. It was to be preserved but, like so many other aircraft, was destroyed post-Armistice. P.C.10 overall with clear doped fabric lower surfaces. White individual and squadron markings. Serial has a thin white outline on red and blue rudder stripes.

5. D.H.5 serial A9165, aircraft "A" of No. 24 Squadron, RFC. This aircraft was credited with five victories. Lt. Wollett was credited with four enemy aircraft between 23 July and 17 August 1917. A fifth victory was credited to Capt. Beanlands. As an aircraft from the first order, actually the third production aircraft, it features the early cowl. Cowl and fuselage panels are natural metal. All fabric upper surfaces were P.C.10. Clear doped lower surfaces. Individual letter in white as is the squadron insignia of a single white bar ahead of the fuselage cockade. Note the application of the serial to the rudder.

De Havilland D.H.6

6. D.H.6 serial C7863 of 530 Flight, Coastal Patrol Unit, based near Dublin. This was an operational aircraft used for anti-submarine patrols. Early broad chord elevators, narrow chord later wings rigged with backstagger. Standard color scheme of grey ply and metal panels; P.C.10 doped fabric upper surfaces and fuselage sides; clear doped lower fabric surfaces. The tailplane and rudder are painted white, possibly as an aid to visibility in the event of ditching. A thin white band was carried on the rear fuselage. Note the "Lift Here" markings, four bladed propeller, and flat windscreen.

7. D.H.6 serial C7617 of No. 203 Training Depot Station, Manston. Early production aircraft manufactured by Grahame-White Aviation Co. Standard color scheme of grey ply and metal panels; P.C.10 doped fabric upper surfaces and fuselage sides; clear doped lower fabric surfaces. Large white individual numeral "18." Note that cockades are carried on both upper and lower surfaces of the all wings. The wheels have been reversed to give a wider wheel track.

8. D.H.6 serial A9738 of No. 121 Squadron. This was the second RAF squadron for American volunteers. Formed in January 1918, the squadron was never operational, crews being transferred to make up losses in existing squadrons. This D.H.6 is in the standard color scheme of grey ply and metal panels; P.C.10 doped fabric upper surfaces and fuselage sides; clear doped lower fabric surfaces. In addition to the white fuselage band, the nose is painted white with wheel covers halved in a dark color (red) and white. Note white outline to serial.

9. D.H.6 serial A9576 of No. 13 Training Squadron shown as it

appeared at Yatesbury. Standard color scheme of grey ply and metal panels; P.C.10 doped fabric upper surfaces and fuselage sides; clear doped lower fabric surfaces. White triangle on fuselage (unit marking?) and white "1" on fin of early style (broad elevators) tailplane.

10. D.H.6 serial C5194 of No. 250 Squadron, Coastal Patrol Unit, based at Padstow. Standard color scheme of grey ply and metal panels; P.C.10 doped fabric upper surfaces and fuselage sides; clear doped lower fabric surfaces. This is a Curtiss OX-5 powered D.H.6 with narrow chord wings rigged with back stagger. Late style narrow chord rudder and elevators. White fuselage marking is probably the unit marking, but this is not confirmed. Note unusual position of serial. Wheel covers thought to be red, white, and blue.

De Havilland D.H.9

11. D.H.9 serial B7620, aircraft "A" of No. 211 Squadron, RAF was forced down and interned in the Netherlands on 27 June 1918. Allotted to No. 211 Squadron about April 1918, it was returning from a raid on Bruges when it was hit by AA fire. The crew of Capt. J. Gray and Lt. J. Comeford landed near Cadzand where they were interned. This aircraft carried the overall P.C.10 scheme with clear doped fabric under surfaces. The rear fuselage was liberally marked with white stripes. The inlets to the nose cowl were painted to represent eyes and a mouth was painted on the lower cowl to complete the face. The individual aircraft letter "A" was repeated under the nose. Note the position of the fuselage cockade. This aircraft was operated by the Dutch with most of the color scheme retained. It was used to train Netherlands East Indies Army pilots.

12. D.H.9 serial C6297, aircraft "3" of No. 144 Squadron. This aircraft force landed in Turkish territory after a raid on Deraa on 16 September 1918, probably due to the unreliable Puma engine. The aircraft force landed in an area where it was impossible for another aircraft to land and the crew of 2/Lt. T. Gitshaw and Lt. C. Thomas walked out. In order to prevent its use by the enemy, the wreck was bombed. P.C.10 overall with clear doped fabric under surfaces. Individual letter and stripe in white. Note that the white individual numeral and white stripe were repeated on the rear fuselage decking with the "3" readable from the rear.

13. D.H.9 serial B7609, aircraft "4A" of No. 203 Training Depot Squadron, Manston, UK, March 1918. This Westland-built D.H.9 was delivered to the RNAS D.H.4 squadron at Manston on 23 February 1918. (This training squadron became No. 203 TDS on 14 July.) It crashed on a test flight on 20 November 1918 and was written off. Grey metal nose panels with the rest of the airframe P.C.10 with clear varnished lower surfaces. Individual aircraft identification in white.

14. D.H.9 serial D7204, aircraft "J" of No. 211 Squadron. P.C.10 overall scheme with clear doped fabric under surfaces. Individual letter "J" has a black shadow outline. This had a thin white outline. Wheel discs white with a red spiral. This aircraft also landed in the Netherlands after a raid on the Bruges docks, crash landing in Zeeland on 24 August 1918.

15. D.H.9 serial D8302, aircraft "N" of No. 108 Squadron, RAF. This aircraft was brought down by AA fire near Ostende on 18 August 1918. The crew of Capt. R. Ingram and 2/Lt. A. Wyncoll became POWs. Metal panels and ply areas of fuselage are grey while upper surface of wings, fuselage fabric covered areas, etc., are doped with P.C.10, Note the white individual aircraft letter is repeated in white under the nose.

De Havilland D.H.9A

16. D.H.9A serial F1010, aircraft "C" of No. 110 Squadron, RAF. No. 110 was the first squadron to be equipped with the D.H.9A. All the original aircraft were presented by the Nizam of Hyderabad as was this aircraft. Allotted to the squadron at Kenley, this Westland built D.H.9A went with the squadron to France on 31 August. On 25 September its crew of Capt. A. Inglis and 2/Lt. Bodley, were credited with a Fokker D.VII shot down "out of control." However, on 6 October the aircraft was shot down in a raid on Kaiserslautern-Pirmanses and the two became POWs. Surviving the war, the aircraft was placed on display in the Berlin *Deutsche Luftfahrt Sammlung*. After this museum was destroyed by RAF bombing during WW2 the surviving airframes were moved to safety in Poland. In 1977 the remains were exchanged for a Spitfire and removed to the RAF Museum Hendon. With new wings and a completely restored fuselage the aircraft is on display at Hendon, the last surviving D.H.9A. Overall P.C.10 with clear varnished fabric lower surfaces. Note "C" on forward fuselage and inscription in white. Note also the white wall tires and position of white fuselage serial. Serial on rudder is black with white outline.

17. D.H.9A serial A1-1 of RAAF. This aircraft incorporated most of the modifications made by the RAAF to keep their Imperial Gift D.H.9A biplanes in service. Flown by Flt. Lt. Charles Eaton during search for the missing Westland Widgeon *Kookaburra*, it suffered engine failure and crashed in outback Australia. Note the oleo undercarriage, large windows to the gunner's cockpit, and footboard stowed on the fuselage top decking. V.84 aluminium doped fabric covered surfaces, with metal panels and ply areas grey. Black serial. RAAF roundels were the same as British but the blue was lighter, closer to that used in World War I.

18. D.H.9A serial E8538, "E-15" of the US Marine Corps Northern Bombing Group. P.C.10 fabric covered upper surfaces and clear doped lower fabric covered surfaces. Ply and metal panels in grey. British serials (in white on fuselage and black on rudder) were retained. USMC serial in black on rudder. Marine Corps insignia on fuselage.

19. D.H.9A serial E9709, aircraft "S" of No. 10 Training Depot Squadron, RAF, Harling Road, late 1918. No. 10 TDS was one of the few training squadrons to receive the D.H.9A before the end of the war. Delivered to the squadron on 28 September 1918, it was transferred to No. 57 Squadron by April 1919. Metal and ply fuselage panels are grey while fabric upper surfaces are P.C.10. White stripe, letter "S", and serial on fuselage. Note black serial to rudder has thin white outline.

20. D.H.9A serial F1019, aircraft "C" of No. 99 Squadron, RAF, France, 1918. No. 99 was one of the few squadrons to receive the D.H.9A before the Armistice. Lt. R.L. McK Barbour and Capt. M.E.M. Wright were credited with a Fokker D.VII shot down on 21 September 1918 in this aircraft. It crashed on landing in January 1919, and returned to 2 Air Depot, presumably to be written off. Natural metal cowl panels. Rest of upper surfaces in P.C.10. Lower fabric covered surfaces clear doped. Westland logo to rear fuselage in black. Fuselage serial and individual aircraft letter in white. Serial to rudder black with white outline.

21. D.H.9A serial F1048, aircraft "T" of No. 205 Squadron, RAF, Verviers, France, February 1919. The squadron was not fully equipped with the D.H.9A until early October 1918. This aircraft was delivered to the squadron in September and returned to the UK in February 1919. It later served with No. 57 Squadron. Natural metal cowl panels. Rest of aircraft P.C.10 upper surfaces with clear doped fabric lower surfaces. Individual letter "T" and fuselage serial in white. Westland logo to rear fuselage in black. White outline to black rudder serial. Red wheel covers.

22. D.H.9A serial E8553, aircraft "N" of No. 155 Squadron, RAF, Chingford, UK. Allotted to the squadron during its mobilization in late 1918. Grey metal and ply panels. Fabric upper surfaces P.C.10 with fabric lower surfaces clear doped. White individual letter and fuselage serial. Black serial to rudder. White and blue (?) wheel covers.

23. D.H.9A serial F1626, aircraft "F" of No. 552 Flight (No. 221 Squadron), Petrovsk, Russia, 1919. Natural metal cowl panels. Rest of upper surfaces in P.C.10. Lower fabric covered surfaces clear doped. Westland logo to rear fuselage in black. The radiator cowl is a dark (red?) color. Fuselage serial and individual aircraft letter in white. Serial to rudder black with white outline. Sold to the anti-Bolshevik government, its eventual fate is unknown.

24. D.H.9A serial E802, aircraft "A" of No. 30 Squadron, RAF, Iraq, circa late 1920s. This Whitehead-built D.H.9A is finished in the standard post-war scheme of V.84 aluminium overall. The wingtips were red on upper and lower surfaces to aid location in the event of a forced landing. The tailplane was also red. The squadron insignia was a palm tree through the numerals "30" in white (or possibly aluminium). The fuselage band was outlined in red; however, the actual color of the band on E802 has not been determined. This band, which extended over the fuselage top decking, was the usual place for the squadron insignia, however this was moved to the fin at a later stage and the aircraft, as depicted, may have been in an intermediate stage. Note the red "A" to the nose and red wheel covers. The black fuselage serial was carried on a white panel. The black rudder serial was outlined in white. Note the desert modifications; the extra wheel carried on the nose, the window at the gunner's cockpit, the carry-all on the bomb carriers, and the enlarged gravity tank. At this point in its career E802 did not have the extra radiator mounted. E802 served from 1924 until 1930, spending its last two years with No. 30 Squadron.

De Havilland D.H.10
25. (Inside back cover) D.H.10 F1868, the third D.H.10 to arrive in France, reaching Marquise on 10 November 1918. Delivered to No. 104 Squadron, RAF, it later saw service with No. 216 Squadron. Standard 1918 bomber scheme with grey ply and P.C.10 upper fabric surfaces, clear doped fabric undersurfaces.

De Havilland D.H.11
26. (Back cover) Prototype D.H.11 Oxford in standard P.C.10 upper surfaces with clear doped lower surfaces. National markings in usual positions.

Selected Bibliography
Henshaw, T., *The Sky Their Battlefield*, Grubb Street, London, 1995.
Cutlack, F.M., *The Australian Flying Corps in the Western and Eastern Theatres of War 1914-1918*, Angus and Robertson, Sydney, 1953.
Sutherland, L.W., *Aces and Kings*, Angus and Robertson, 1935.

2. D.H.5, No. 68 (Australian) Squadron

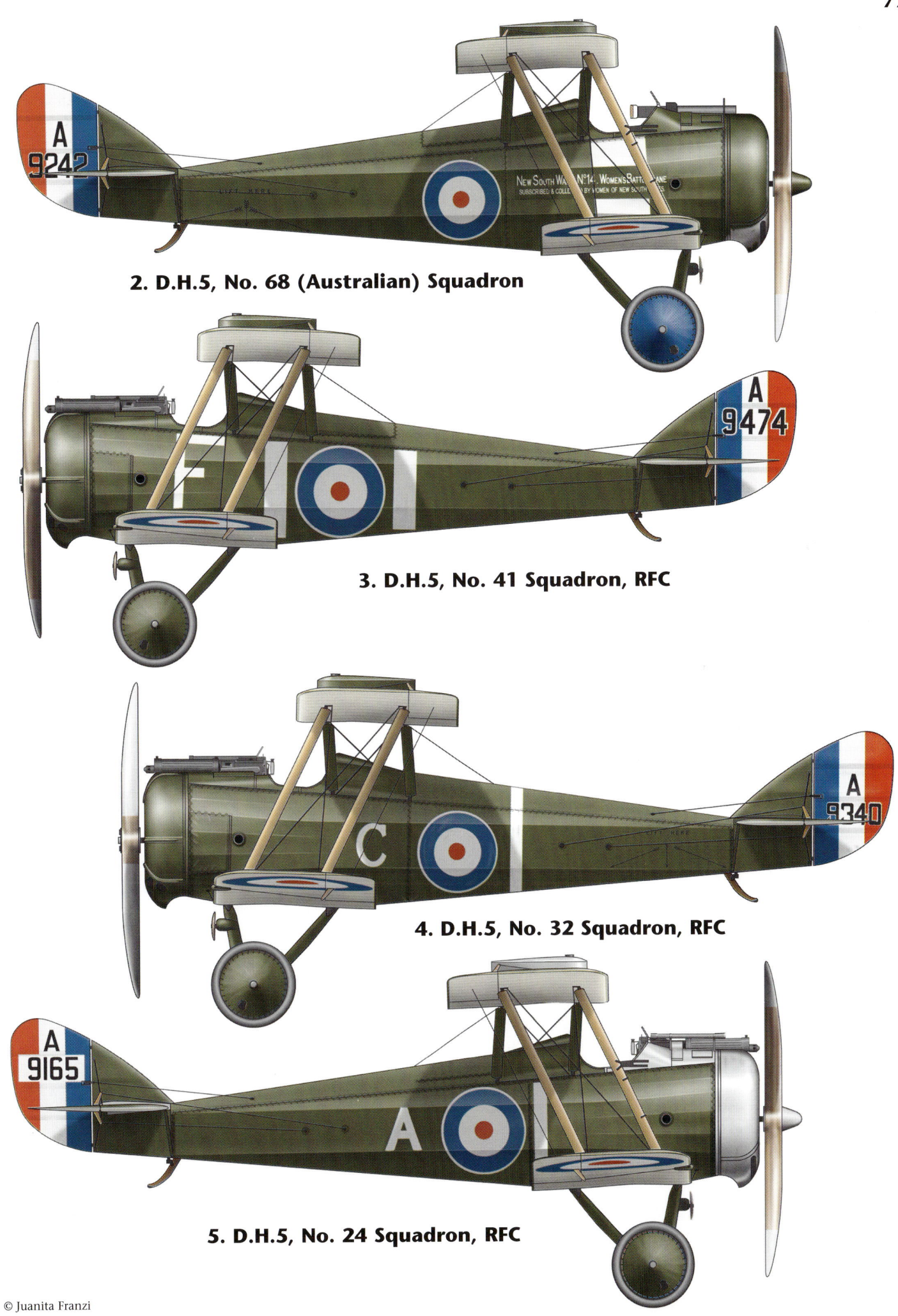

3. D.H.5, No. 41 Squadron, RFC

4. D.H.5, No. 32 Squadron, RFC

5. D.H.5, No. 24 Squadron, RFC

© Juanita Franzi

6. D.H.6, 530 Flight, RFC

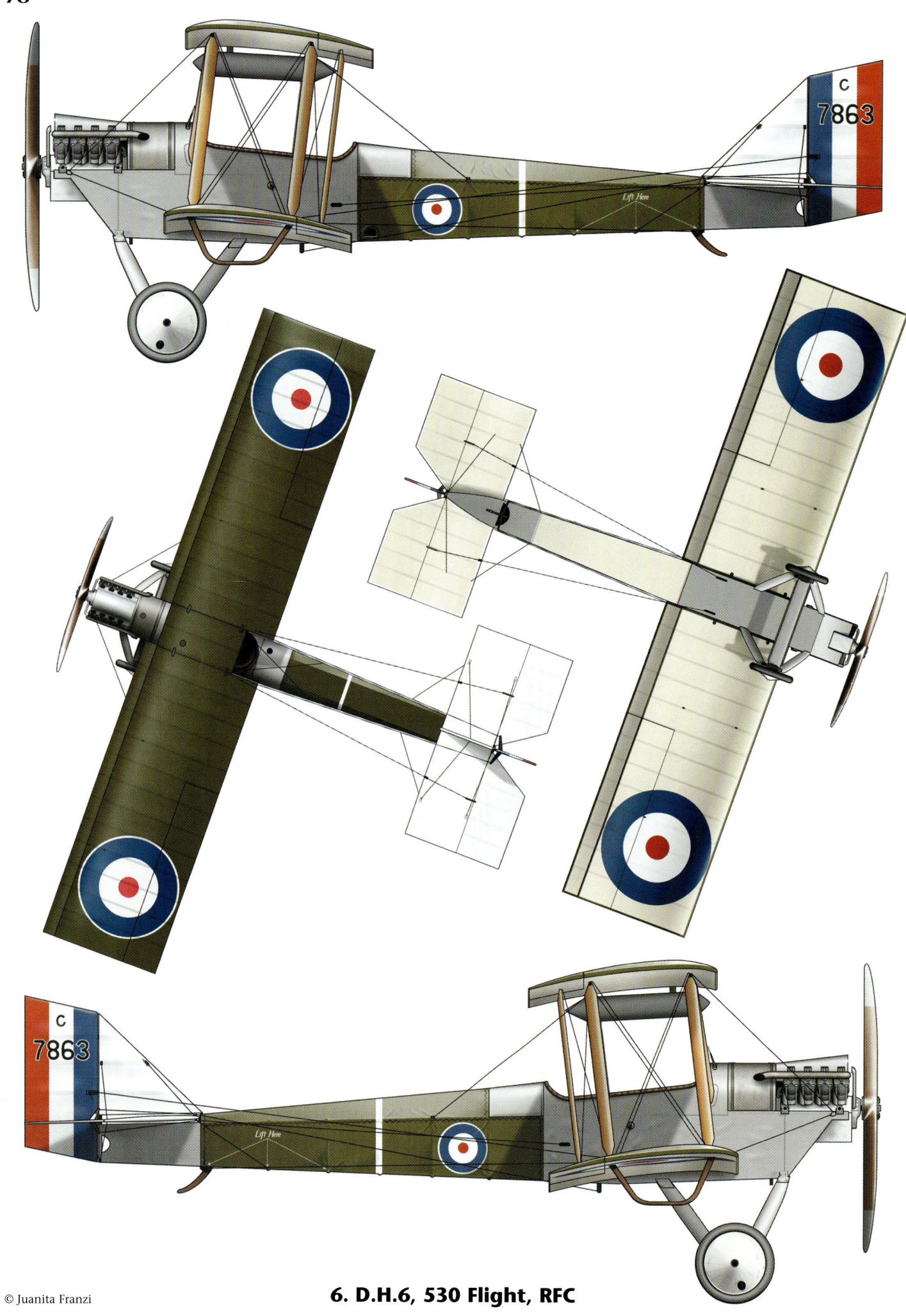

79

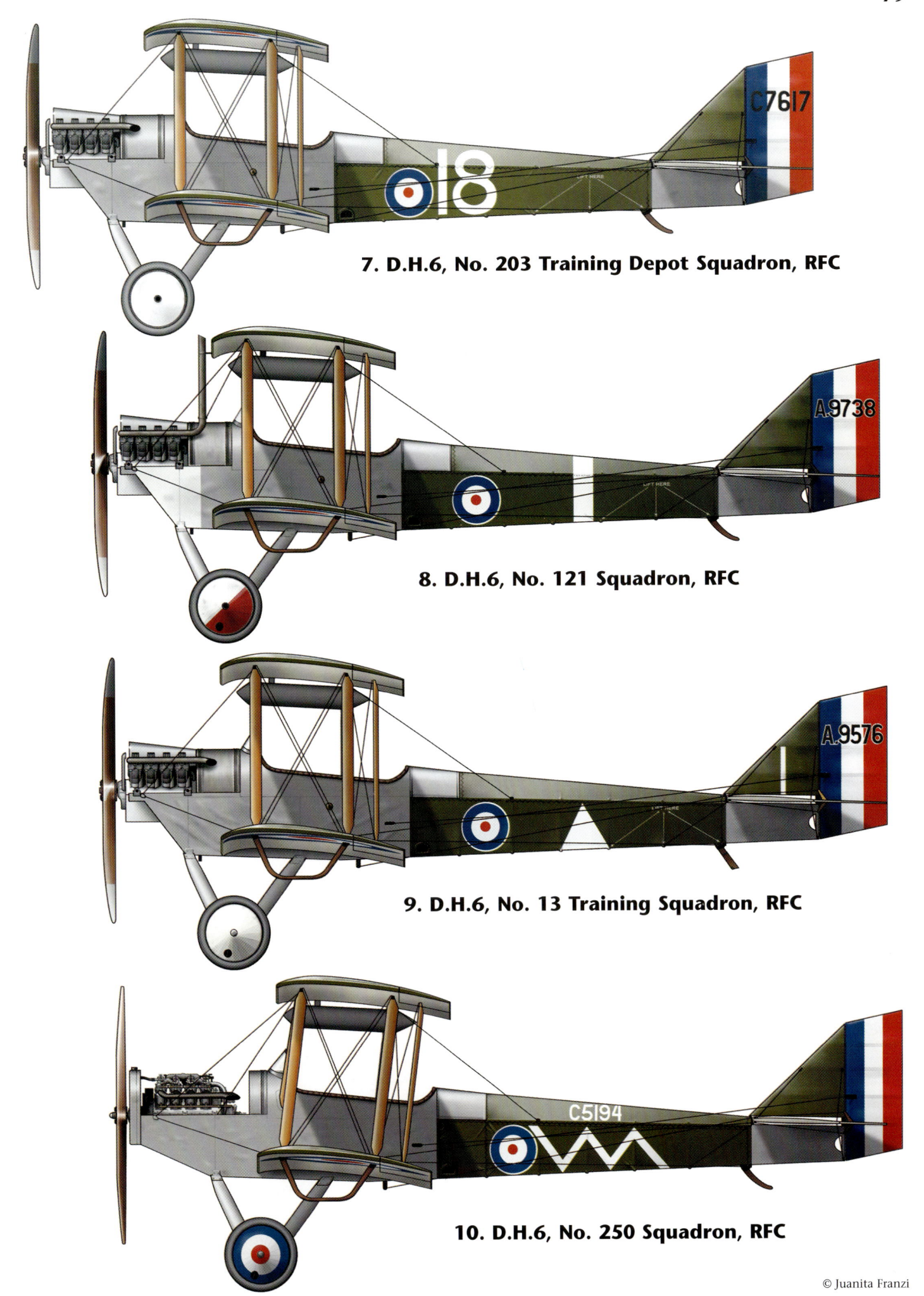

7. D.H.6, No. 203 Training Depot Squadron, RFC

8. D.H.6, No. 121 Squadron, RFC

9. D.H.6, No. 13 Training Squadron, RFC

10. D.H.6, No. 250 Squadron, RFC

© Juanita Franzi

11. D.H.9, No. 211 Squadron, RAF

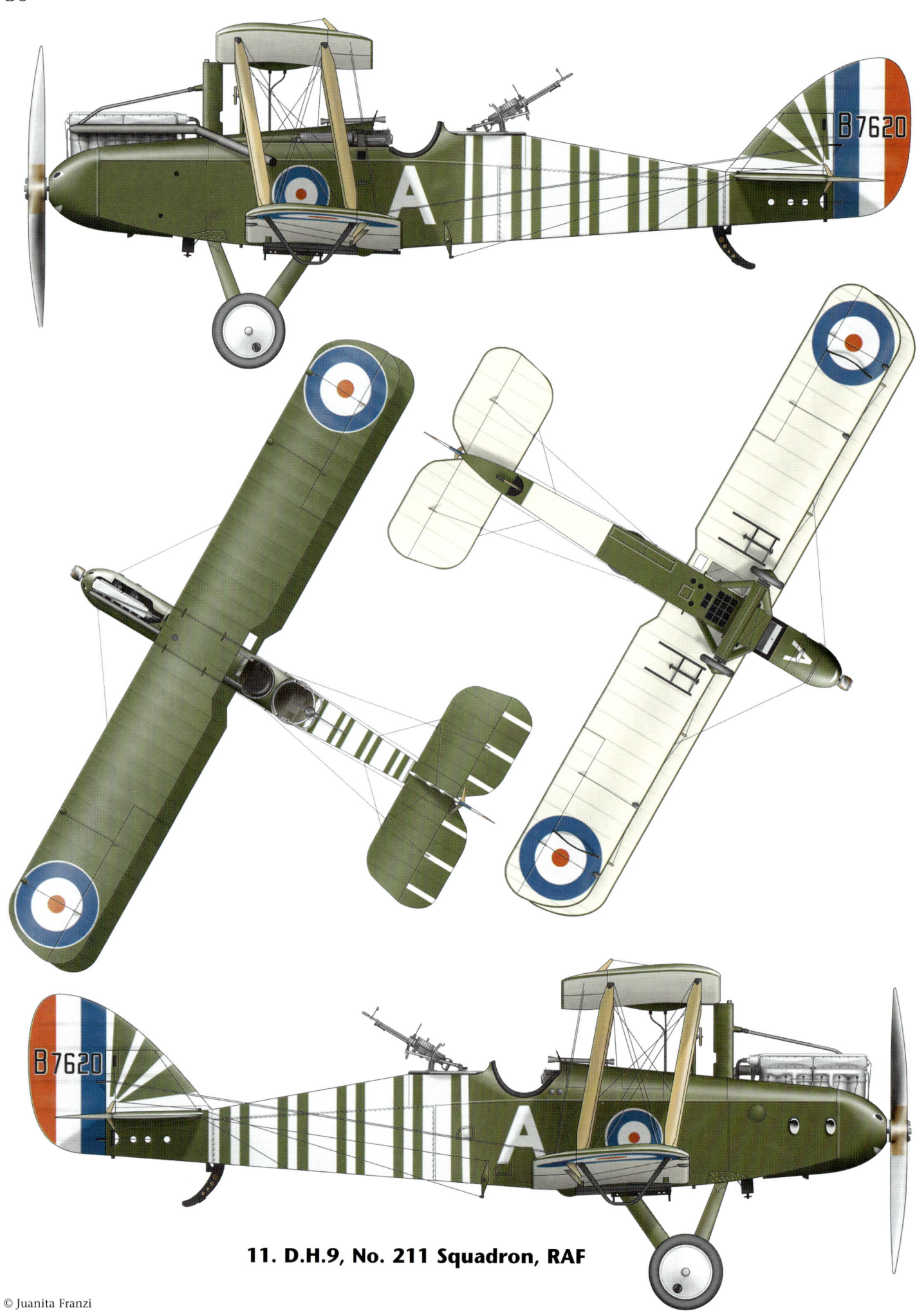

81

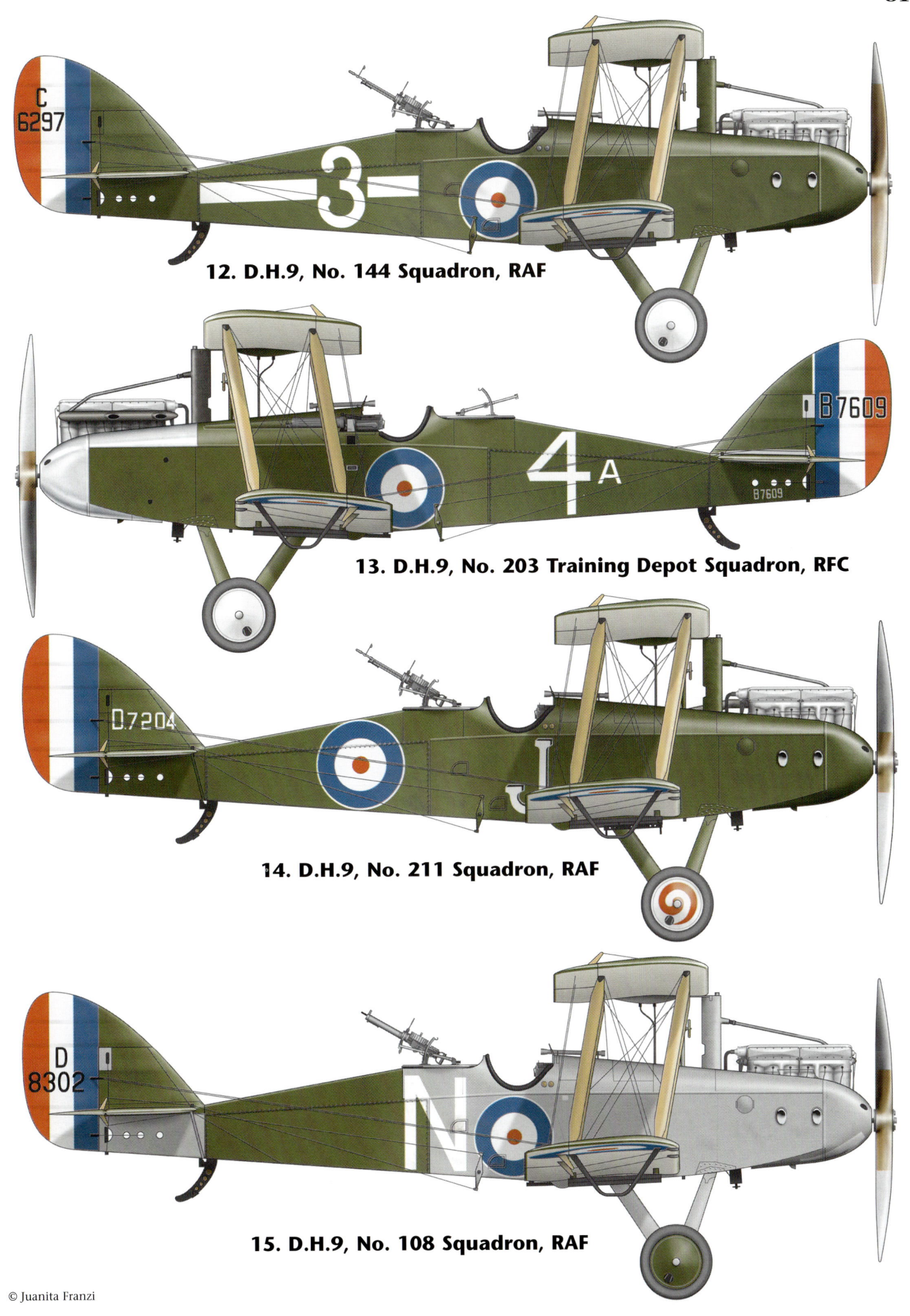

12. D.H.9, No. 144 Squadron, RAF

13. D.H.9, No. 203 Training Depot Squadron, RFC

14. D.H.9, No. 211 Squadron, RAF

15. D.H.9, No. 108 Squadron, RAF

© Juanita Franzi

16. D.H.9A, No. 110 Squadron, RAF

17. D.H.9A, RAAF

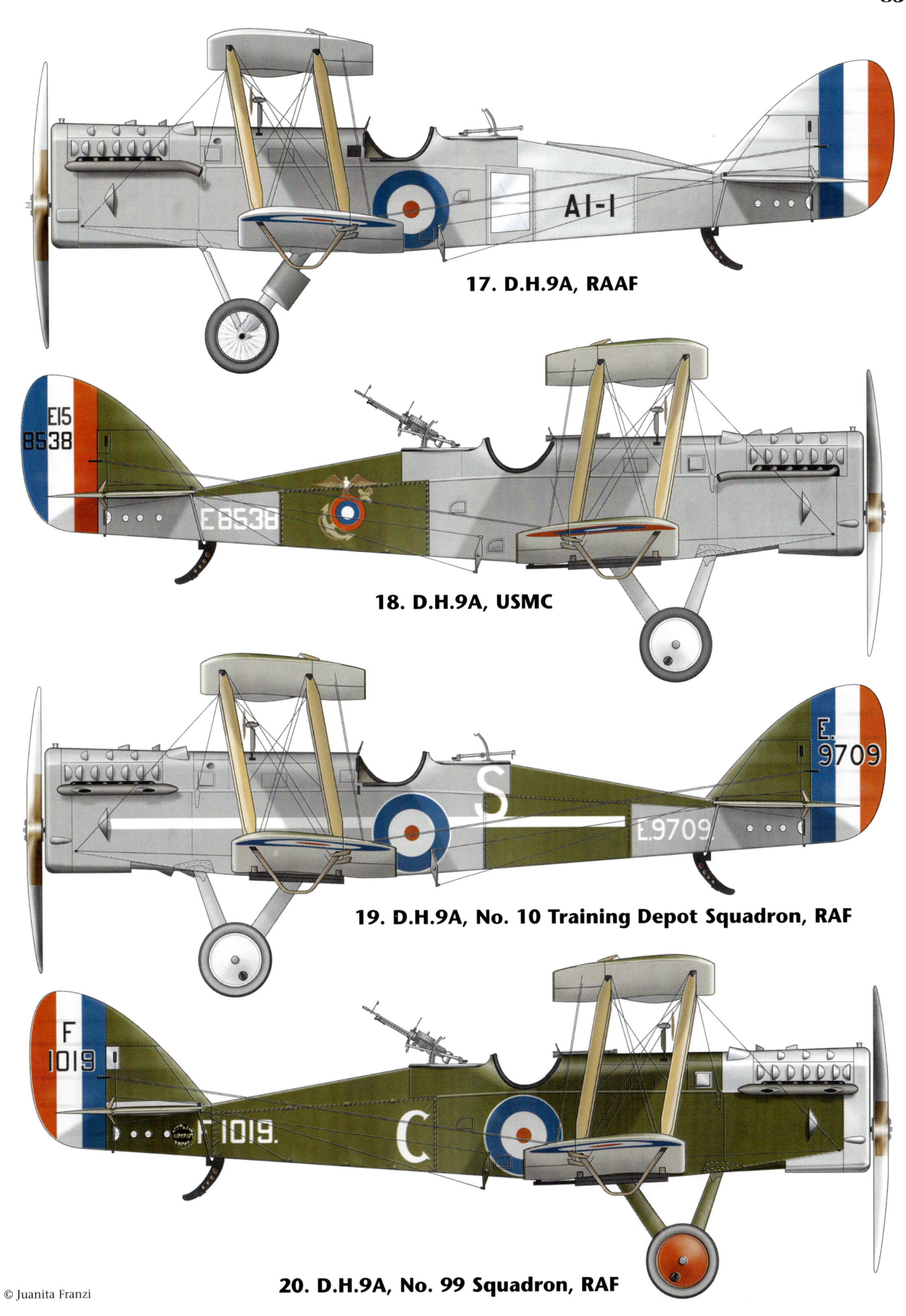

18. D.H.9A, USMC

19. D.H.9A, No. 10 Training Depot Squadron, RAF

20. D.H.9A, No. 99 Squadron, RAF

© Juanita Franzi

21. D.H.9A, No. 205 Squadron, RAF

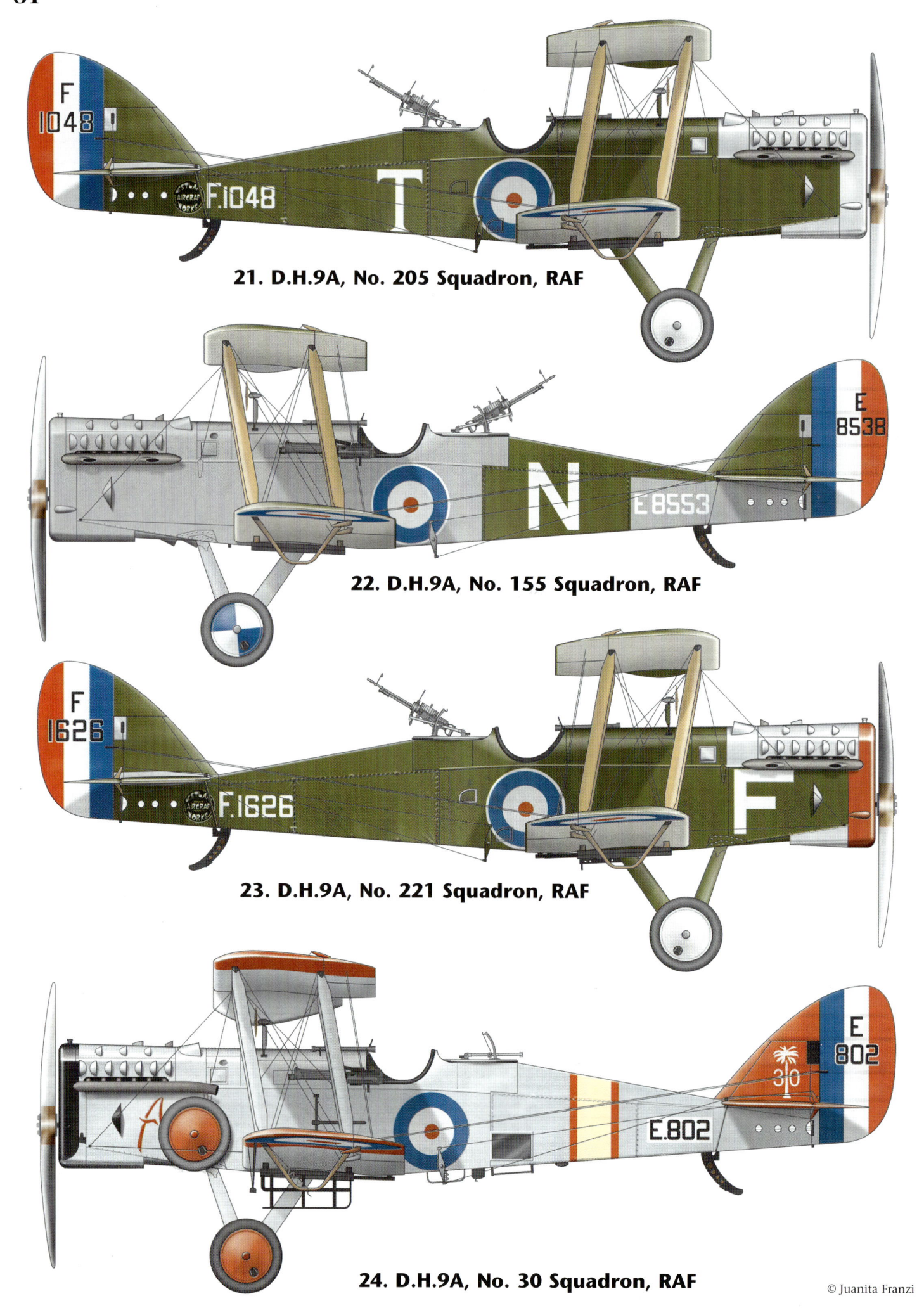

22. D.H.9A, No. 155 Squadron, RAF

23. D.H.9A, No. 221 Squadron, RAF

24. D.H.9A, No. 30 Squadron, RAF

© Juanita Franzi

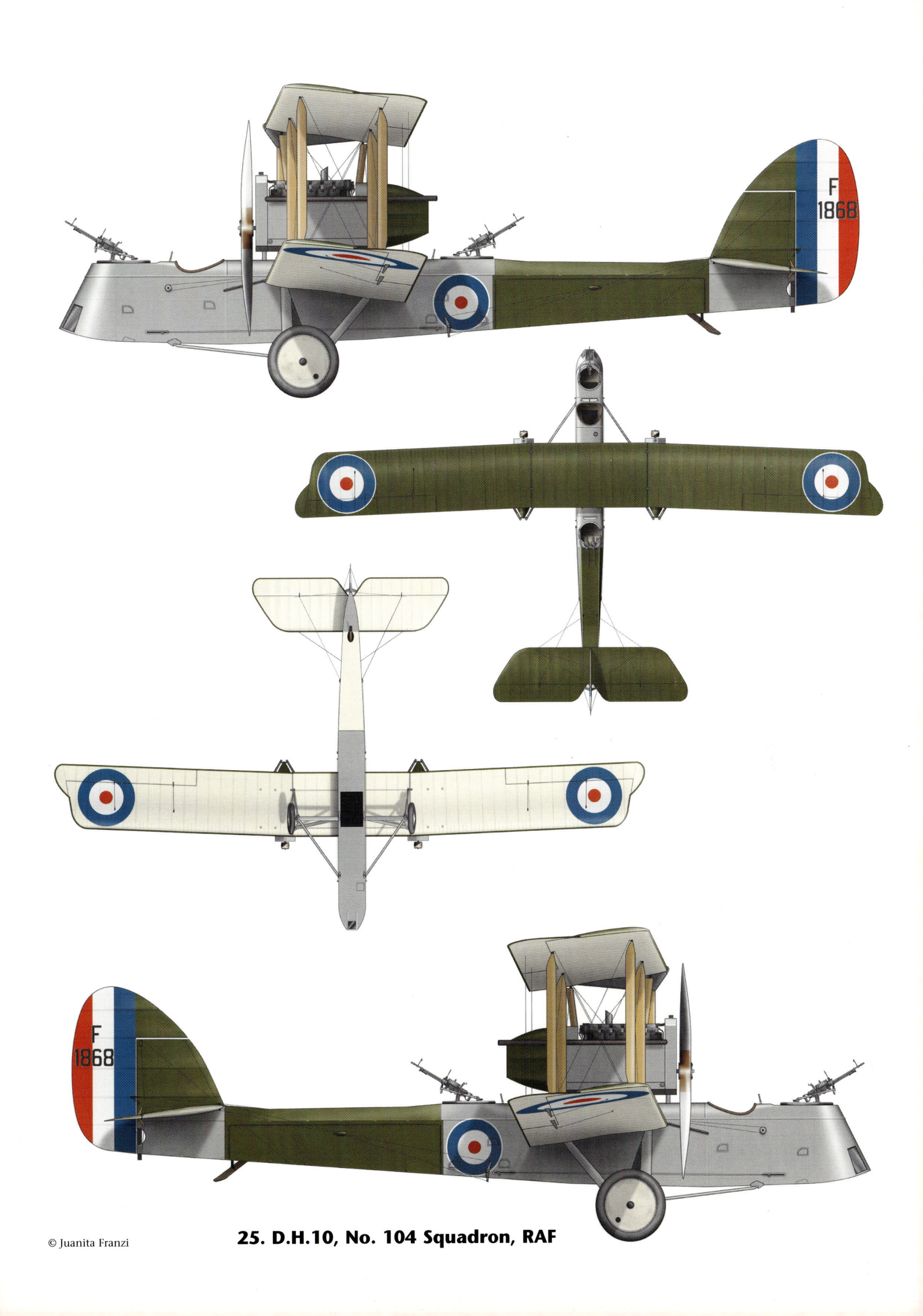

25. D.H.10, No. 104 Squadron, RAF

26. D.H.11 Prototype

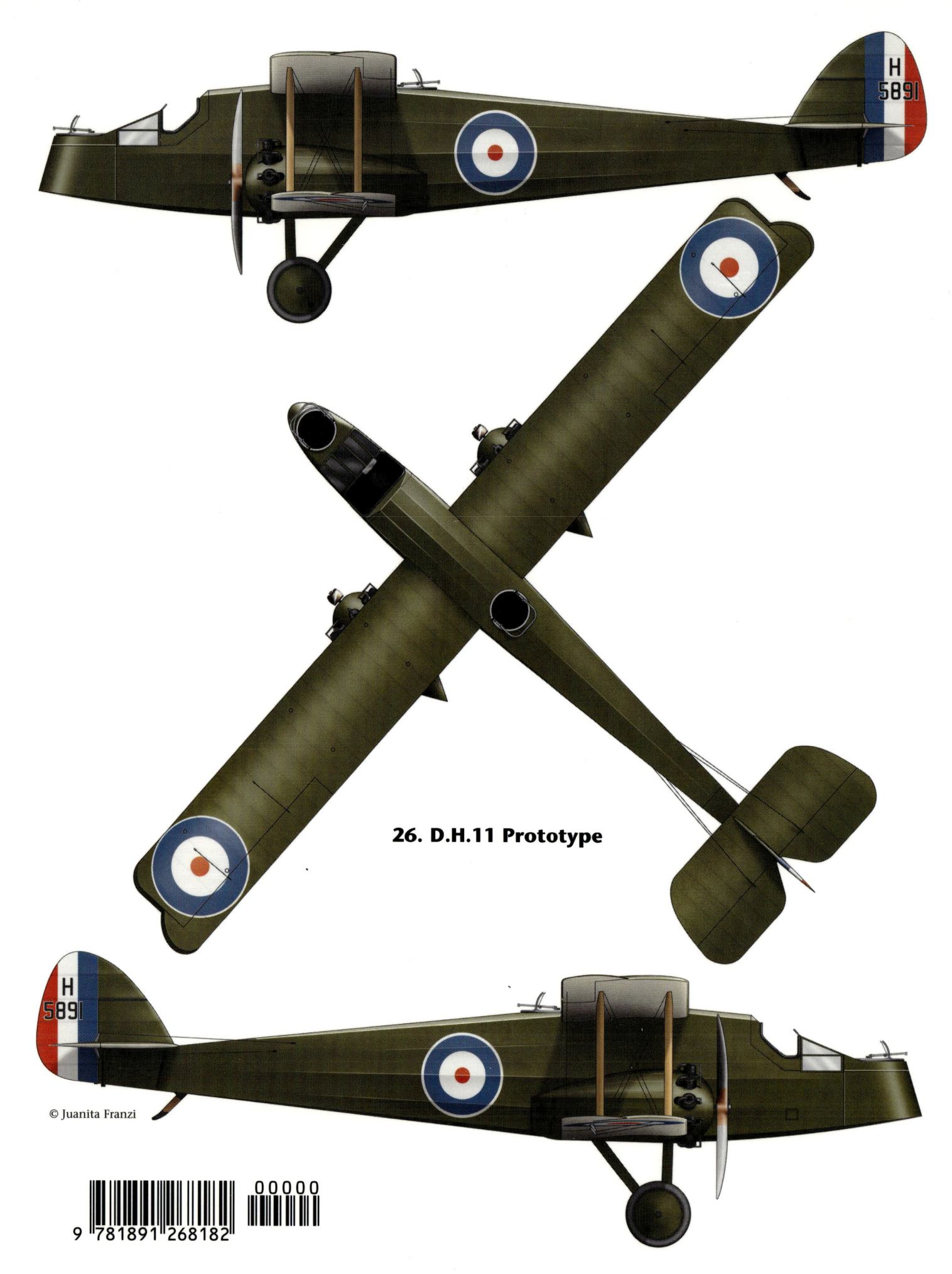

© Juanita Franzi

9 781891 268182

FMP 2006

A Flying Machines Press Book • ISBN 1-891268-18-X

Visit our Web site at: www.flying-machines.com